GETTING RESULTS:
THE SIX DISCIPLINES OF
PERFORMANCE-BASED PROJECT MANAGEMENT

ANN COSTELLO
GREGORY A. GARRETT

ACQUISITION
SOLUTIONS

. Wolters Kluwer
Law & Business

AUSTIN   BOSTON   CHICAGO   NEW YORK   THE NETHERLANDS

Editorial Director: Aaron M. Broaddus
Cover & Interior Design and Layout: Craig L. Arritola

Copyright Notice

Notice of Trademarks

Product No.: 0-5015-400

ISBN: 978-0-8080-1818-6

CONTENTS

PREFACE

The United States Federal Government spends more than $400 Billion annually to acquire products, services, and integrated solutions to provide for our nation's defense, space exploration, energy, education, health care, justice, commerce, transportation systems, and much more. Most of the money expended by the U.S Federal Government is spent to purchase large, complex systems, integrated information technology solutions, and a wide variety of professional services. In an effort to get the most for our U.S. tax-payers dollars, the Office of Management and Budget (OMB), and the Office of Federal Procurement Policy (OFPP), have required the expanded use of Performance-Based Acquisition (PBA), improved project management, and focused contract administration.

The principal authors and editors of this book, Ann Costello and Gregory A. Garrett are both highly respected and experienced industry business executives with extensive knowledge of U.S. Federal Government acquisition management, commercial contracting, strategic sourcing, and project management best practices. In addition, Costello and Garrett are considered two of the foremost thought leaders in the topic of Performance-Based Business. Costello, the Principal Author of the *7 Steps to Performance Based Services Acquisition* developed this in partnership with numerous U.S. Government agencies. Garrett is the best-selling and award-winning author of 10 books and 70+ published articles, including: *Performance-Based Acquisition: Pathways to Excellence; Managing Complex Outsourced Projects; Leadership: Building High Performance Buying and Selling Teams;* and *U.S. Military Program Management: Lessons Learned & Best Practices.*

Costello and Garrett have collaborated with five of their highly distinguished colleagues currently working together at Acquisition Solutions, Inc. (ASI). The group is composed of executives, consultants, and former U.S. Government, military, and/or private sector industry leaders.

This ASI team of acquisition all-stars from U.S. Government and industry has written an insightful and compelling book about what it takes to Get Results, by applying the six-disciplines of Performance-Based Project Management (PBPM):

1) Cultural Transformation
2) Strategic Linkage
3) Governance
4) Communications
5) Risk Management
6) Performance Management

Further, the authors have included an outstanding discussion of how to manage a multisector workforce and proven effective practices to capture, transfer, and reuse organizational knowledge to improve project results.

The book concludes with a thought provoking discussion of the future Acquisition Workforce and what roles, responsibilities, and competencies will be needed to achieve high performance results in the management of high technology, complex products, services, and integrated solutions to support the growing needs of the United States of America.

Said simply, this book is outstanding and a must read for everyone involved in Getting Results, especially for those who care about improving the business of government.

Sincerely,

Lenn Vincent, RADM, U.S. Navy (Ret.)
Industry Chair
Defense Acquisition University

DEDICATION

Gregory A. Garrett

I would like to dedicate this book to my lovely and talented wife Carolyn for her friendship, patience, love, support, and the three greatest gifts—our children Christopher, Scott, and Jennifer.

Ann Costello

This book is dedicated with heartfelt thanks to the men and women of Acquisition Solutions, whose talent, innovativeness, integrity, teamwork, excellence, optimism, and energy continually impress and inspire me. It is also dedicated, with the deepest love and gratitude, to my husband George and my sons, Kevin and Ryan, whose support has enabled to me to pursue my career goals and dreams.

ACKNOWLEDGMENTS

The coauthors and principal editors Ann Costello and Gregory A. Garrett would like to deeply thank the following people for their valuable contributions and teamwork to make this book become a reality!

ACQUISITION SOLUTIONS, INC.

Contributing Authors

Anne Reed

John Gaeta

Shirl Nelson

Shaw Cohe

Bill Kaplan

Administrative Support

Justin Goodman

Barbara Hanson

Julie Oliver

WOLTERS KLUWER LAW & BUSINESS — CCH

Professional Support

Aaron Broadus, Esq. – Editorial Director

Sharon Kube – Product Manager

ABOUT THE AUTHOR

Ann Costello

Ann Costello is Acquisition Solutions' Managing Partner, responsible as the thought leader for our knowledge management offerings. Ann is renowned for her skills in organizing information and her clear and concise writing abilities. She pioneered the firm's corporate research and publications division—now known as the Acquisition Solutions Research Institute™—and wrote most of the early Acquisition Directions™ *Advisories, Updates,* alerts, and best-practice reports.

Ann, whose career spans public service, self-employment, and corporate management, is an expert in Federal acquisition policy and management and an innovator in information presentation. She works both internally and externally to produce cutting-edge, multi-dimensional knowledge management tools. For example, Ann was the Content Master and Information Architect during the development stage of the Acquisition Directions™ knowledge management portal. Ann also served as project manager and industry partner on an interagency team (led by the Department of Commerce, with the participation of the Departments of Treasury, Agriculture and Defense, the General Services Administration, and the Office of Federal Procurement Policy) to develop the Web-based guide and knowledge center, *Seven Steps to Performance-Based Acquisition.* Of the site, Professors Ralph Nash and John Cibinic said in their February 2002 Report, "For ease of use and for providing a vast amount of information, we rate the guide a '10'." Ann also developed Acquisition Solutions' Web-based knowledge management tool—the Acquisition Solutions Virtual Knowledge Center on Using Federal Supply Service Multiple Award Schedule Contracts™.

Ann has hundreds of publications to her credit. One of many notable achievements is her primary authorship of the GSA-published *Guide to Planning, Acquiring, and Managing Information Technology Systems,* which was called by the then-OFPP Administrator the "best document of its type." She has also authored a number of detailed implementation guides and handbooks on subjects such as *Feasibility, Alternatives, and Cost/Benefit Analysis; Strategies for Successful Federal Supply Schedule and Blanket Purchase*

Agreement Acquisitions; and *Making a Business Case for Information Technology Acquisitions.*

Ann served as a primary subject matter expert to the Federal Acquisition Institute and used her Federal knowledge base to develop new performance measurement processes for states' public benefits information systems. She managed major programs, including the GSA's Brooks Act delegations and procurement management reviews programs.

Ann holds a Bachelor of Science degree in Business Education from Radford College, and she has completed graduate-level study at The George Washington University.

Gregory A. Garrett

Gregory A. Garrett (Gregg) is an international educator, best-selling and award-winning author, and highly respected business leader. Currently, he serves as Chief Operating Officer for Acquisition Solutions, Inc. where he leads the consulting engagements and training for all U.S. Federal Government agencies, including the Department of Defense, Homeland Security, Treasury, State, Veteran Affairs, Commerce, Agriculture, Justice, Labor, Education, Interior, Energy, Health and Human Services, Small Business Administration, National Science Foundation, Equal Employment Opportunity Commission, NASA, and others.

Prior to Acquisition Solutions, Gregg served as an executive with Lucent Technologies in positions including: Chief Compliance Officer, U.S. Federal Government Programs; Chairman, Project Management Center of Excellence; Vice President, Program Management, North America, Wireless Commercial Accounts; and Chairman, Lucent Technologies Project Management Leadership Council, representing more than 2,000 Lucent project managers globally.

Prior to Lucent, Gregg served as a Partner and Executive Director of Global Business for ESI International, where he led the sales, marketing, negotiation, and implementation of global business management, bid/proposal management, government contracting, and project management training and consulting programs for numerous Fortune 100 multinational corporations. He has served as a lecturer for The George Washington University Law

School and the School of Business and Public Management. He has personally taught and consulted in bid/proposal management, contract, and project management to more than 25,000 people from over 40 countries.

Formerly, Gregg served as a highly decorated military officer for the United States Air Force, where he was awarded more than 17 medals, badges, and citations. He completed his active duty military career as an Acquisition Action Officer, in the Colonel's Group Headquarters USAF, the Pentagon. He was the youngest Division Chief and Professor of Contracting Management at the Air Force Institute of Technology, where he both led and taught advanced courses in purchasing, contract administration, and program management to more than 5,000 people from the Department of Defense, NASA, and industry. Previously, he was the youngest Procurement Contracting Officer for the USAF at the Aeronautical Systems Center, where he led more than 50 multi-million-dollar negotiations and managed the contract administration of more than $15 billion in contracts for major weapon systems. He served as a Program Manager at the Space Systems Center, where he managed a $300 million space communications project. He is DAWIA Level III certified in contracting and program management.

Gregg holds a Bachelor's degree in Chemistry/Engineering Physics from Miami University and a Master of Science degree in Systems Management from the University of Southern California. He also successfully completed Executive Education in Advanced Project Management at Stanford University. He is a Certified Project Management Professional of the Project Management Institute and has received the prestigious PMI Eric Jenett Project Management Excellence Award and the PMI David I. Cleland Project Management Literature Award. He is a Certified Professional Contracts Manager and a Fellow and member of the Board of Advisors of the National Contract Management Association. He has received the NCMA National Achievement Award, the National Education Award in contract management, the Charles J. Delaney Memorial Award for Contract Management Literature, the Blanche Witte Memorial Award.

A prolific writer, Gregg has authored 12 books, including: *Managing Contracts for Peak Performance* (NCMA 1991), *World-Class Contracting* (Fourth Edition, CCH 2006), *The Capture Management*

Life-Cycle: Winning More Business (CCH 2003), *Managing Complex Outsourced Projects* (CCH 2004), *Contract Negotiations* (CCH 2005), *Contract Management Organizational Assessment Tools* (NCMA 2005), *Performance-Based Acquisition: Pathways to Excellence* (NCMA 2005), *Leadership: Building High-Performance Buying & Selling Teams* (NCMA 2006), *U.S. Military Program Management: Lessons Learned & Best Practices* (Management Concepts 2007), *Solicitations, Bids/ Proposals, & Source Selection: Building a Winning Contract* (CCH 2007), and *Cost Estimating & Contract Pricing* (CCH 2008). He is also the principal author and editor of the Federal Acquisition Action Packs series published by Management Concepts. Further, he has authored more than 70 published articles on bid/proposal management, contracting, project management, supply chain management, and leadership.

INTRODUCTION

This book focuses on what it takes to achieve great business results in the complex world of U.S. Federal Government contracts and projects. Specifically, the book addresses: the nature of the blended (multisector) workforce challenges and opportunities, the need for knowledge management throughout the acquisition life cycle, and the mandate to provide effective program/project management in an environment of performance-based acquisition. The book provides a comprehensive discussion of the six integrated disciplines of Performance-Based Project Management (PBPM), including:

- Cultural Transformation
- Strategic Linkage
- Governance
- Communications
- Risk Management
- Performance Management

Further, the book provides 100+ proven best practices, tools, techniques, and more than 12 case studies from both U.S. government agencies and industry. The book concludes with a brief discussion of the Future Acquisition Workforce and what it will take to get great results!

We hope you will find this book informative and enjoyable!

Sincerely,

Ann Costello
Gregory A. Garrett

GETTING RESULTS: THE SIX DISCIPLINES OF PERFORMANCE-BASED PROJECT MANAGEMENT

By Anne Reed and Gregory A. Garrett

INTRODUCTION

The world in which we live is changing at an ever-increasing rate. Likewise, business paradigms are shifting significantly for organizations and for each individual. Downsizing, outsourcing, horizontal supply-chain relationships, globalization, and increased pressures to achieve high-performance results are affecting nearly everyone, especially business leaders. Within the buying and selling communities, organizations are struggling to advance from classic "stove-pipe" structures to talent-based integrated project teams. With the power of tremendous communication technologies at their fingertips, individuals are now challenged to overcome the overwhelming daily influx of e-mails, voice mails, and instant messages to establish focused priorities and ensure flawless execution.

Indeed, the mantra of business today is flawless execution, whether buying or selling products and services. Every organization in both the public and private business sectors is trying to increase its velocity of business (the speed in which it meet its customers' needs), while reducing costs and simultaneously ensuring it is providing quality products and/or services to meet or exceed its customers' expectations. There is no question the performance bar has been raised in all aspects of business. Some customers have such incredibly high performance expectations that they may seem impossible to achieve. Yet, what appears impossible today often becomes the performance norm of tomorrow.

In both the public and private business sectors, we are witnessing increased use of outsourcing and performance-based contracts. Today, more than ever, buyers are requiring customized, integrated solutions to their business challenges. Often these customized solutions require multiple suppliers, each with multiple functions, to team together, to seamlessly deliver solutions composed of hardware, software, and professional services. All of these forces are transforming the marketplace—and placing greater risks on both parties.

Sellers are being asked to agree to increasingly demanding performance-based contracts, with specific product performance requirements and service-level agreements, often with significant penalties for failure to perform. Buyers, too, are taking greater risks, because more and more they are outsourcing their capabilities

ONE

to other parties in full or in part (through creation of multisector workforces), thus losing some direct control over their success or failure. Business is becoming more electronically enabled, less personal, more regulated, and highly interdependent. Thus, the world we live in and the forces of emerging technologies, growing customer demands, and increased government regulations are forming a new performance-based supply environment.

In the public business sector, composed of federal, state, and local government agencies, the adoption of more commercial buying practices clearly is part of a quest to reduce acquisition cycle time, reduce expenses, and increase customer satisfaction. Similarly, the private business sector is moving to create better, more defined and effectively implemented business processes. Especially in the wake of the U.S. Sarbanes-Oxley Act, private-sector companies are struggling to increase speed to market, improve profitability, and increase customer confidence and loyalty.

The public and private business sectors are in many ways converging through technology and regulations, resulting in two distinct trends: (1) the increased use of electronic tools to rapidly procure simple off-the-shelf products and services, and (2) the increased use of large, complex performance-based deals. While the vast majority of business transactions are procurements of simple commercial off-the-shelf products and services, these transactions typically comprise just 20 percent of the spend value of most U.S. government agencies and commercial companies. In contrast, the large complex performance-based contracts, while fewer in number, typically comprise 80 percent or more of the spend value of most U.S. government agencies and commercial companies.

Thus, the focus of this book is on getting results through the application of the six disciplines of performance-based project management, as described by Acquisition Solutions, Inc., in managing the large, complex, performance-based contracts widely used by U.S. government agencies. Said simply, performance-based contracts are good, but excellence in execution is better! Many U.S. government agencies are struggling to deliver the best bang for the buck, but they are using the wrong measure. Too often, they focus their time and talent on selecting the right source or supplier or contractor and awarding the contract. Then it's on to the next award. Too often, they fail to focus critical resources on managing the deal/

contract/project to ensure excellence in execution—meeting or exceeding mission requirements. This book describes the techniques and disciplines that should be applied preaward and postaward to ensure effective performance-based project management.

WHAT IS PERFORMANCE-BASED PROJECT MANAGEMENT?

As conceived by Acquisition Solutions, performance-based project management (PBPM) encompasses six critical disciplines supported by repeatable, cyclical processes. The six disciplines of performance-based project management are:

1. _Cultural Transformation_ – Proactively manage the organizational and cultural changes integral to the success of the project.
2. _Strategic Linkage_ – Provide a consistent vision throughout the organization, making sure the desired project results reflect organizational strategic goals.
3. _Governance_ – Establish roles, responsibilities, and decision-making authorities for project implementation.
4. _Communications_ – Identify the content, medium, and frequency of project information flow to all stakeholders.
5. _Risk Management_ – Identify, assess, monitor, and manage project risks.
6. _Performance Management_ – Plan, analyze, and report status—cost, schedule, and performance—on a regularly scheduled basis during project execution.

Performance-based project management is an execution-focused application of the concepts of performance-based _acquisition._ Incorporated into the implementation are tools and techniques for project planning, defining, scheduling, baselining, communicating, training, and reporting. PBPM _is a complete, strategic approach to ensuring that the business of the government, delivered through contract, is managed based on performance._ The following sections address the six disciplines of PBPM and how they are applied to ensure project success.

Discipline 1: Cultural Transformation

Cultural transformation encompasses identifying how an organization and its people need to adapt to make a project successful, as well as shepherding that organization through acceptance and institutionalization of the new processes and behaviors.

ONE

How do you prepare for a cultural transformation?

First, you must be prepared to invest in creating the right project environment and processes. This environment must promote and encourage measuring performance and delivering results; must motivate individuals, teams, and the entire organization to execute their authority and fulfill their responsibilities; must stimulate government and contractor employees to perform work and achieve the desired results; must inspire shared results; and must ensure repeatable processes.

To create this type of environment to support performance-based project management, you will need:

- **Executive Leadership** – Establish an environment of empowerment, responsibility, and active participation at the top that will cascade down through the organization. The leader identified as executive sponsor of the project must be someone who is respected by the organization faced with change.
- **Trust** – Establish good working relationships and partnerships among all project participants, including all functional areas and various business partners.
- **Clarity** – Establish a clear picture of what people are expected to do and achieve. Use the organization's mission, individual roles and responsibilities, definitions for success, performance expectations, and performance reporting to create unambiguous expectations.
- **Accountability** – Establish, as appropriate, individual and group responsibility for actions, so there is a personal interest in the outcomes.
- **Consistency** – Establish consistent policies, procedures, and resources so stakeholders and participants understand the organization's commitment to improving the business environment.
- **Planning** – Develop a formal change management plan that addresses organizational readiness, gap analysis, and training (or other specific requirements) that will be needed to prepare for the cultural transformation.
- **Preparation** – Help the workforce adapt to this new way of doing business by using techniques such as coaching, mentoring, training, facilitated sessions, and exercises in teamwork. Remember that your participants will range from early adopters to those who prefer a "wait and see" approach.
- **Reinforcement** – Keep sending the transformation message; the old and familiar way of doing business will have a stronger

pull than the new. This causes people to backslide into old habits and process, sometimes without even being aware they are doing so.

Chapter 4 discusses how to accomplish a cultural transformation in either a U.S. government agency or industry to align the workforce to achieve mission results via PBPM.

Case Study — Defense Contract Management Agency

The Defense Contract Management Agency (DCMA) is a Department of Defense (DoD) organization composed of thousands of people. Its blended workforce includes active-duty military personnel representing every branch of the U.S. armed forces, civilians, and contractors, all dedicated to providing quality contract management support for DoD and various civilian government agencies, including the National Aeronautics and Space Administration (NASA). DCMA's workforce performs a wide range of professional services, including preaward surveys, quality assurance, change management, contract administration, in-plant inspections, engineering/technical evaluation support, and much more. Its personnel are deployed worldwide in support of DoD.

For the past few years, DCMA has been on a cultural transformation journey, evolving from rigid rules, regulations, manuals, and highly detailed specifications to a more flexible set of business guidelines for each functional area. It has developed its new business guidelines on the basis of proven best practices, adapted from both internal and external benchmarking. Large, complex organizations do not change overnight; it took DCMA years to evolve the thinking and business practices of its thousands of workers, via education, hands-on training, coaching, and follow-on assessments. But, according to its recent customer surveys, the cultural transformation has made DCMA more customer focused and results-based.

Discipline 2: Strategic Linkage

Strategic linkage provides a consistent vision throughout the organization, vertically and horizontally, making sure the intended results from the contract and related project reflect and support organizational strategic goals and are achieved during performance-based project management. This linkage is both organizational and personal.

How do you align an organization "strategically"
to achieve project goals?

One of the keys to initial project budgetary approval is alignment with the organization's mission and strategic goals. Acquisitions should be guided by a statement of objectives (SOO) developed to define success in words that fit the critical success factors identified in the organization's strategic plan. This will establish the framework for what is important to measure. The discipline of strategic linkage does not end with the initial contract award, but continues throughout project execution. It is possible to have a project that is performing on schedule, within budget, and delivering exactly what was envisioned—but that is out of alignment because of unanticipated changes in policy or strategic direction. You need to establish processes to ensure there are checkpoints to revalidate the strategic linkage and take corrective action where appropriate.

In addition, individual performance expectations should be linked to project and organizational expectations. Individuals' performance plans should be directly linked to the overall success of the project—which is linked to the success of the organization's mission. Strategic linkage between the organizational stakeholders, the integrated project team, the subcontractors/suppliers, and the individual is discussed in much more detail in chapter 5.

Case Study — Honda

Honda is deeply committed to satisfying the needs of every one of its stakeholder groups. It delivers industry-leading quality to its customers, motivates and leverages its people, creates mutually beneficial relationships with suppliers, and consistently delivers superior value for shareholders. At the same time, it remains engaged with surrounding communities and generally stays on the right side of governments and regulators. How does Honda do it?

The key is that beyond simply maintaining a controlled balance between stakeholders, Honda has implemented many practices and programs that create synergies between these groups. These include its executive compensation plans, employee motivation tactics, the REACH (Recognizing Efforts of Associates Contributing at Honda) program of rewards, and the Best Partner initiative, as well as the company's corporate structure.

To align employee and investor interests, Honda has implemented many performance-based pay initiatives for its staff. These practices enable the company to reward results and innovation without excessive pay packages. Honda's top management makes less money than that of its competitors. By closely tying executive compensation to company performance, Honda ensures a greater transparency and fairness for investors and sends the right message to the rest of its workforce. If Honda is doing well and customers are happy, employees benefit (through bonuses) along with investors.

The Honda culture continually rewards good ideas and empowers employees to look for more efficient ways to work. Under the REACH program, the company is able to gather the best ideas from employees for continually improving quality, innovation, and efficiency, giving individuals "kaizen awards" for their successful ideas. Honda also offers awards for detecting defects in product quality and safety hazards in the plants. The company has given many cars to participants in the REACH program over the years, as well as cash rewards.

Because Honda relies heavily on vendors to supply most of the components of its products, it has rightly recognized that suppliers—when treated as partners—can enable positive synergies across the enterprise. Collaborating with suppliers helps improve quality and ensures that costs are tightly controlled, both of which relate directly to Honda's customer value proposition. High-quality, reliable products at affordable prices are possible and sustainable only if the whole supply chain is efficient. In working with its supply chain, Honda successfully balances a partnership mind-set with tough standards on price and quality. As noted in the 2003 North American Automotive Tier 1 Supplier Study, "Honda and Toyota, and to some extent Nissan, recognize that they can pressure their suppliers for considerable price reductions and quality improvements and still have good supplier working relations. It all comes down to how you work with people that determines whether or not you get the best performance from them."

An important component in Honda's supplier strategy is the Best Partner program. A dedicated team works with suppliers to help them achieve the high standards and target costs that Honda demands. This creates a collaborative and synergistic relationship that benefits both the supplier and Honda. This is apparent in another finding of the above-mentioned study: "In terms of which OEM

ONE

provides them the best opportunity to make an acceptable profit margin on their business, suppliers rate Toyota and Honda highest, followed by Nissan, Daimler Chrysler, Ford, and GM."

Discipline 3: Governance

Governance encompasses the establishment of roles, responsibilities, and decision-making authorities necessary for successful project execution.

How do you "govern" the process?

A governance model should be developed carefully and with deliberation. Many well-conceived projects fail because decisions are not made in a timely manner or are made by the "wrong" people. Costly misunderstandings ensue and trust is eroded when it is unclear who has the authority to make what decisions. The governance process is a key aspect of accountability—on the part of both government and industry.

The governance model includes an organizational framework along with processes for ensuring the objectives of a project continue to be met after contract award. Tailored contract governance structures help identify and address issues and challenges early, as well as establish ways for resolving issues and disputes. For a project to flow smoothly and successfully, the governance process must involve representatives from all the stakeholder groups, with clearly assigned roles, responsibilities, and reporting relationships among them.

A robust governance model will, for example:
- Clearly establish roles and responsibilities of key stakeholders (such as the executive sponsor, program manager (government and industry), the government's contracting officer's technical representative (COTR) and contracting officer, and the customers).
- Identify and create charters for advisory and decision-making bodies (such as the Executive Steering Committee, Program Management Office, Change Control Board, and integrated project teams).
- Define decision-making and related authorities for key individuals (for example, who recommends, who approves, and who signs?).
- Identify a dispute resolution process.

To be effective, the governance model should recognize that there are some decision frameworks that are unique to the government, some that are unique to the industry partner(s), and some that require joint ownership. Chapter 6 provides a more detailed discussion of all aspects of governance and how it is vital to project success.

Charter Components	
Purpose	**Membership**
Scope	Roles & Responsibilities
Authority	Meeting Quorums
Organization	Operating Guidelines
Key Milestones	

Discipline 4: Communications

Communications involves identifying the content, medium, and frequency of project information flow, as well as the organizational elements the information is intended to support.

How should you communicate to effectively support project success?

Every project requires collaboration—communications before, during, and after contract award can make the difference between success and failure. Communication must occur on many levels. You must first define your project organization's communications objectives. You also must identify the needs of key internal and external audiences critical to achieving the objectives of the project, as well as preferred or appropriate communication styles. All stakeholders will want to be regularly informed about the status of the project.

Having defined the project communications objectives and audience, you should then develop a communication plan. The plan should identify the processes needed to ensure the timely and appropriate generation, collection, dissemination, and storage of project information. Communications mediums may range from formal briefings to newsletters, video teleconferences, or websites. Regardless of the vehicle chosen to distribute information, messages should be tailored to the audience and must be honest, clear, consistent, frequent, open, and complete (see the chart below).

Communication Principles
Honesty is the only policy. Enough said.
Segment the audiences. Acknowledge the fact that different audiences have different needs.
Use multiple channels. Use a variety of vehicles to ensure maximum penetration and receptivity.
Communicate clearly. Make the content of the message clear and specific. Anticipate questions to create clear messages.
Communicate frequently. Apply the "Rule of Seven." You must say the same thing seven times in seven different ways before anybody will believe it.
Anticipate barriers. Structure communication to overcome barriers.
Allow feedback. Provide mechanisms for feedback and respond to all feedback.
Connect people. Provide communications that foster a sense of "connectedness" between people.
Tailor messages. Tailor specific messages to meet the needs and preferences of stakeholders.
Cascade messages. Use agency leadership to send the messages to all.
Adapted from: Go-Live Communication Plan, State of Oklahoma CORE at: http://www.youroklahoma.com/coreoklahoma/go-live2.pdf.

Remember that training is another form of communications. When new processes or procedures are introduced—for example, a new way of accessing a help desk, a new web-based data entry format, or a new decision-making framework for process approval—people need to be trained so they are prepared for program execution. Team-based training is a helpful tool in assisting team members to adapt from a compliance-based relationship with their business partners to a performance-based relationship. Chapter 7 discusses the art of project and organizational communications in more detail.

Case Study: Wal-Mart

Wal-Mart has mastered the art of clear, concise, and compelling communication to help its employees worldwide focus on the organization's core mission: "Excellent success at the stores." One of Wal-Mart's proven best practices is its top management, buyers, and key marketing personnel meeting every Saturday morning. The "Saturday Morning Meeting" provides a quick but thorough review of revenues and profit, what is going well, and what needs to be fixed. It often includes an inspirational speaker or a customer who delivers a compelling message!

Shortly after the weekly Saturday Morning Meeting, a summary of all of the key information is broadcast from Wal-Mart headquarters in Bentonville, Arkansas, to all of the company's stores via the chain's six-channel satellite system. In addition, the satellite communication system enables Wal-Mart store managers to talk with each other as frequently each day as they need.

Wal-Mart's real-time communication system, combined with its culture of continuous improvement and focus on lowering costs and improving customer satisfaction, has enabled it to become the world's leading retailer.

Discipline 5: Risk Management

Risk management is treated in performance-based project management as processes that identify, assess, monitor, mitigate, and manage risk.

How can we manage risk?

Every contract and project carries a certain element of risk. Some require exceedingly complex or untried technical solutions; some necessitate large-scale collaboration, which can inhibit decision making. Occasionally, the project is dependent on the incorporation of a component that could be subject to delays in availability; sometimes there are unusual cost constraints or situations where required skill sets are in scarce supply. It is important to acknowledge risk and plan how to address it. You should start with a risk management plan. The risk management plan outlines how you will approach and plan risk management activities for a project. It should identify risks, outline risk mitigation actions, and provide guidance on monitoring progress in mitigating risks.

How can we mitigate identified risks?

While the actual mitigation tactics used depend on the unique circumstances of a particular risk, there are some generic strategies that can be used. These include:

- **Avoidance**. Changes a "project plan to eliminate a risk or condition or to protect a project's objectives from its impact. Although a project team can never eliminate all risk events, some specific risks may be avoided."[1]
- **Transference**. Shifts "the consequences of risk to a third party together with ownership of the response. Transferring a risk simply gives another party responsibility for its management; it does not eliminate it."[2]
- **Mitigation**. Reduces "the probability and consequences of an adverse risk event to an acceptable threshold. Taking early ac-

tion to reduce the probability of a risk occurring or its impact on the project is more effective than trying to repair consequences after it has occurred."[3]

- **Acceptance**. Accepts that risk may occur. When a "project team decides it is unable to identify a suitable response strategy, the team may develop contingency plans, should the risk occur. Active acceptance may include developing a contingency plan to execute, should a risk occur. Passive acceptance requires no action, leaving the project team to deal with the risks as they occur."[4]

> Risk is treated in performance-based project management as processes that identify, assess, monitor, mitigate, and manage risk.

Chapter 8 discusses risk management in much more detail.

Case Study — Boeing Integrated Defense Systems

As companies become more successful in dealing with complex project challenges, risk management typically becomes a more structured and integrated process that is performed continuously throughout the business life cycle. Such is the case at Boeing Integrated Defense Systems, where designing, manufacturing, and delivering an aircraft can take years and require a multibillion-dollar investment. Typically, Boeing evaluates the following risk categories and develops detailed risk mitigation strategies and actions to improve its business case by reducing or eliminating potential negative aspects. Risk categories include:

- Financial – Up-front funding and payback period based upon number of planes sold.
- Market – Forecasting customers' expectations on cost, configuration, and amenities based on the 30- to 40-year life of a plane.
- Technical – Forecasting technology and its impact on cost, safety, reliability, and maintainability.
- Production – Supply-chain management of a large number of subcontractors with no impact on cost, schedule, quality, or safety.

Discipline 6: Performance Management

Performance management encompasses project planning, scheduling, budgeting, analyzing, monitoring, and reporting on status on a regularly scheduled basis throughout project execution.

Are there success factors for measuring performance?

There are seven critical success factors that need attention both preaward and postaward. First, focus on project results. Remember that in a performance-based contract, you are driving toward a programmatic objective, not compliance with a contract. One of your evaluation criteria in a performance-based contract award should be the quality of the performance metrics proposed by the contractor. If so, you have a solid foundation.

Second, be prepared to negotiate if the proposed performance metrics are not well stated or clearly defined in the contract. Both the government and contractor should have an interest in establishing well-defined metrics that meet the government's objectives and require the contractor to "stretch" for a good evaluation and past performance rating.

Third, reach agreement with the contractor on the performance metrics *before* contract award and be ready to monitor progress toward achieving the project results at the onset of the contract. This is critical.

Fourth, designate who from the government and who from the contractor will serve on the performance management team; develop the methodology to be followed if this has not been done.

Fifth, track only the most essential performance measures and ensure that the cost of the measure is worth the gain. Tracking too many or too costly performance measures is burdensome and could compromise success. Quantity does not mean quality.

Sixth, establish an expectation that performance measures will be reevaluated over time, setting a framework for continuous improvement. Modify performance measures and measurement processes to respond to changing needs and priorities.

Seventh, check each measure against quality standards. Good project performance measures are: (1) valid and objective, being based on reliable and accurate data, sources, and methods; (2) cost effective in terms of gathering and processing information; (3) understandable and easy for decision-makers and stakeholders to use and act on; and (4) tied to incentives wherever possible. There should always be a clear link between achieving a specified performance target and some form of meaningful compensatory reward or recognition. The better the relationship between the measure, reward, and outcome, the better the performance.

How should you manage performance?

Simply measuring performance is not enough. Measurement carries an obligation to *manage*. If performance improvement is not as great as anticipated, the project management team should take whatever action is feasible and reasonable under the circumstances to improve it. Such action may include determining problem areas, monitoring performance more closely, changing what is being measured to incentivize better performance, reallocating resources, or devising methods for improvement.

In general, manage performance by measuring the efficiency and effectiveness of the organization's and contractor's performance after contract award. You also need to measure the quality of the business outcomes and how efficiently they are achieved.

Contractors should be rewarded when the desired outcomes are achieved—and given disincentives for not achieving established standards and acceptable quality levels. In support of this process, numerous techniques are available to review and track the progress of a project, develop relevant performance measures, and incorporate them into regular project reviews in a systematic and disciplined way. This sets a framework for providing to the right level of management the right information at the right time for decision making. Some common tools include:

- *Quality Assurance Surveillance Plans (QASPs).* A QASP aligns objectives with relevant measures and outlines how progress will be monitored to ensure that the defined performance measures are achieved. It also should define acceptable quality levels, the surveillance methodology, incentives and disincentives.
- *Earned Value Management System (EVMS).* The use of an EVMS is an industry and government standard for defining the processes for baselining, authorizing, taking credit for, tracking, and changing the costs, schedule, and content of a project. It provides the processes for assessing cost and schedule variance based on earned value. Earned value uses the original estimates and progress to date to indicate whether the actual costs incurred are on budget and whether the tasks are ahead or behind the baseline plan. It provides

an integrated view of cost, schedule, and the technical aspect of a project.

- *Service-Level Agreements (SLAs).* SLAs document project goals and objectives, establish task costs and schedules, and set forth performance measures for contract tasks and project-level measures for high-impact, mission-critical tasks. SLAs, which may be part of the QASP or separate, allow the customer and the service provider to identify upfront what services will be provided, how they will be measured, and what happens if the level of service is not delivered as promised.
- *Balanced Scorecard (BSC).* A BSC combines both financial and operational measures into an integrated system of performance indicators. The scorecard provides an enterprise view of an organization's overall performance by integrating financial measures with other key performance indicators related to customer and employee perspectives, internal business processes, and organization growth, learning, and innovation. The intent is to provide a flexible tool that allows organizations to set their goals and track their achievement.

After establishing performance goals and building a performance measurement system, the next step is to put performance data to work by establishing and conducting regular in-progress performance management status reviews that integrate the information for several levels of management. At the same time, "learning before, during, and after" techniques (chapter 3) should be used to identify and document best practices and lessons learned in performance measurement and management for continuous learning and improvement.

> There should always be a clear link between achieving a specified performance target and some form of meaningful compensatory reward or recognition. The better the relationship between the measure, reward, and outcome, the better the performance.

Performance management techniques and systems such as these are not stagnant, but evolving. They help organizations identify what works and what doesn't. They are tools that help track direction and progress toward meeting objectives. Based on information from these techniques and systems, organizations can improve, repair, replace, or discontinue a project. At this point, you will have aligned your project objectives with the organization's business, transformed your organization in preparation for performance-based project management, and established the framework for a disciplined approach to governance, communications, risk management, and performance management. This takes the organization beyond just evaluating the acquisition and contract and into the realm of PBPM. Chapter 9 provides a more detailed discussion of performance management.

Does performance-based project management begin at contract award?

No. That is way too late. Performance-based project management begins early in the process of a performance-based acquisition. The groundwork must be laid well before contract award and cemented soon after with the input from the selected contractor(s).

What are the benefits of performance-based project management?

Performance-based project management completes the performance-based acquisition cycle and is necessary to fulfill its promise. The ultimate benefit of PBPM is that it sets the ground rules and the framework for operating so that U.S. government agencies and contractors can focus on programmatic results—and not get mired in meaningless processes. More specifically, the PBPM methodology:

- Provides a structured approach for focusing on strategic performance objectives.
- Defines the mechanism for accurately reporting project performance to upper management and stakeholders.
- Brings the stakeholders into the acquisition lifecycle—from strategic planning through mission accomplishment and the evaluation of performance.
- Provides a framework to account for results and manage continuous improvement efforts.
- Makes the delivery of project results a joint responsibility between the government and the contractor.

> The ultimate benefit of performance-based project management is that it sets the ground rules and the framework for operating so U.S. government agencies and contractors can focus on project results.

What are the keys to making performance-based project management successful?

PBPM requires an organization and its people to take a different approach to doing business. It requires:

- *Collaboration versus Direction.* Government and contractors must collaborate for mutually beneficial results. Heavy-handed, directed performance removes contractor flexibility and is compliance oriented.

- *Insight versus Oversight.* Reviewing the contractor's progress and methodologies is not nearly as important as understanding the impact of various options. As long as everyone is always well-informed, less oversight is required, allowing a true partnership to develop.
- *Flexibility versus Control.* The roles of the government leaders and managers must change in a performance-based environment. No longer do they review and control the contractor's inputs to processes, products, and deliverables. In a performance-based environment, they should review measures of performance and give contractors the freedom and flexibility to make the trade-off decisions necessary to ensure success.
- *Results versus Compliance Orientation.* In a performance-based environment, the focus needs to be on managing objectives, defining success, and measuring outcomes.
- *Partnership versus Dictatorship.* Traditional contracts are managed to direct the "top-down" implementation of the government's view of what is required to achieve an outcome. In contrast, a performance-based management partnership enables the contractor to deploy its best ideas to accomplish the programmatic results desired by the government.

When government and contractors partner to resolve problems with processes or people, they will both be more focused on the results.

Conclusion

The time has come for agencies and contractors to conduct business differently. Performance-based project management expands the concept of performance-based acquisition, embedding the concepts throughout the acquisition life cycle and driving performance and results throughout an agency's culture and business operations. Performance-based acquisition requires an inherent shift in a government agency's culture, from one focused on control and oversight, compliance and direction, and time-and-materials staff augmentation to one focused on partnership, collaboration, performance, and—ultimately—results. And it requires the six disciplines of performance-based project management to achieve mission results through contractors.

ONE

The next two chapters discuss the challenges and opportunities facing U.S. government agencies and industry and their related workforces in the management and successful execution of complex performance-based contracts, specifically in the areas of managing a multisector workforce and improving knowledge management. The management of the multisector workforce is especially vital to U.S. government agencies to set the proper foundation for implementing performance-based project management.

QUESTIONS TO CONSIDER

1. How effective is your organization's culture in creating high-performance results?

2. Has your organization evolved from functional silos into effective integrated project teams?

3. How well are the strategic goals of your organization aligned with those of the stakeholders, teams, contractors/suppliers, and individuals?

4. How effective is your organization's risk management process?

5. How consistent is your organization in achieving best-in-class results?

Endnotes

1   *A Guide to the Project Management Body of Knowledge, 2000 Edition;* Project Management Institute; Newtown Square, Pennsylvania.

2   Ibid.

3   Ibid.

4   Ibid.

CHAPTER  2

OPTIMIZING PERFORMANCE IN A MULTISECTOR WORKFORCE

By Ann Costello

INTRODUCTION

According to the National Academy of Public Administration (NAPA), "[t]he trend toward using workers who are not part of the federal civil service to carry out federal missions has escalated greatly in recent years."[1] The Professional Services Council agrees, and has for some time, with this assessment. While some held the view that the tightening federal budget pressures would affect the market for professional and technical services, in 2006 the Council was "convinced that the government's growing mission and continued human capital challenges were combining to create a new market dynamic, one that was less directly reflective of overall government budgets and more reflective of the ongoing shift of service delivery from the organic federal workforce to private sector providers."[2]

This trend is certainly the case in terms of the workforce engaged in support of federal acquisition and project management services. In his statement to the Acquisition Advisory Panel in November 2005, Stan Soloway, president of the Professional Services Council, noted that "the widely accepted fact [is] that we have too few acquisition professionals in the right places with the right skills."[3] The situation is likely to continue to erode. Said Soloway:

> The government continues to live with personnel demographics that are wildly out of balance. The inevitable retirement wave presents challenges of its own—there are currently more than two-and-a-half times as many government employees over the age of 55 as there are employees under 30. But those challenges are substantially exacerbated by the government's struggle to attract, or retain, junior- and middle-management professionals who, in a more balanced demographic set, would be in position to repopulate the senior management ranks as the retirement exodus takes place. Unfortunately, the government personnel systems continue to limit agency ability to attract mid-career professionals and the government has not become an "employer of choice" for workers with skills in high demand. And the data show that a modest but steady stream of those critical journeyman professionals now in government are

exiting voluntarily. As such, the government's greatest challenge lies in the likely "bathtub" effect that will occur as the senior workforce departs over the next three to seven years.

For the acquisition workforce, these demographic challenges are just as pronounced.[4]

Thus, the demographic profile of the federal workforce is a powerful contributing force behind the increased reliance on contractors to meet federal mission needs. And, given that it takes five years or more to train and develop an acquisition professional, reliance on the multisector workforce is a necessary reality.

What is the multisector workforce?

"Multisector workforce" is a term adopted by NAPA to "describe the federal reality of a mixture of several distinct types of personnel working to carry out the agency's programs. ... [Use of the term] recognizes that federal, state and local civil servants (whether full- or part-time, temporary or permanent); uniformed personnel; and contractor personnel often work on different elements of program implementation, sometimes in the same workplace, but under substantially different governing laws; different systems for compensation, appointment, discipline, and termination; and different ethical standards."[5] The multisector workforce is sometimes referred to as the "blended workforce."

One example of a large-scale multisector workforce is the Federal Emergency Management Agency (FEMA) Map Modernization program, which cuts across all layers of government. Data are shared with other federal agencies. There are partnerships with state, regional, and local stakeholders that allow partners to choose their level of involvement in mapping tasks such as collecting, updating, and adopting flood data. There are user groups for engineers/surveyors, floodplain managers, homeowners, and insurance professionals and lenders. And, of course, there is the FEMA-contracted national service provider, known as the Mapping On Demand Team.

What advantages are brought by a multisector workforce?

Many benefits have been cited for a multisector workforce, most of which fall under the general category of *workforce management*. They include the ability to:

■ contract out work that is not inherently governmental,

■ focus the federal workforce on core missions,

■ focus contractors on shorter term projects with a limited period of performance,

■ reduce the size of government, and

■ meet workload demands, sometimes in the face of constraining full-time equivalent (FTE) ceilings or hiring freezes.

Sometimes the workforce management argument for a multisector workforce is supplemented by the benefit of *saving money*. More often than not, though, the supplemental argument is *skills* focused, such as to acquire knowledge, skills, and abilities absent or in short supply in the current federal workforce.

Other benefits relate to the general category of *flexibility and speed*, sometimes with a *mission* focus, for example, the ability to:

■ obtain surge capacity and to scale quickly,

■ achieve operational flexibility,

■ adapt quickly to changing mission needs,

■ augment capacity in the event of an emergency,

■ react to increasing mission complexity, and

■ respond quickly to changing and emerging technology with less direct and negative effect on the federal workforce.

With regard to the latter point, to quote NAPA, "the government's rigid, rule-bound civil service system does not facilitate or encourage flexibility in the civil service workforce."[6]

What special considerations or concerns does a multisector workforce introduce?

According to NAPA, "Using multisector workforces requires a dispersion of administrative authority and reduces the level of control that the federal manager has over the process, yet this does not reduce oversight responsibilities or the requirement for delivering results. Moreover, the federal manager is ultimately accountable for program successes and disasters...."[7] To overcome this dispersion of authority and control requires a strong results-based orientation that is expressed in the performance management expectations and processes applied to federal employees through federal merit-based employment principles and practices—and to contract employees though contractual performance objectives and incentives.

There is some concern that members of a multisector team operate under different pay and benefits profiles. Each member of the team is operating as an individual under a contract with his or her employer that sets certain performance and rewards expectations; in addition, contractor employees are responsible for meeting the contractual expectations of the clients they are serving. Avoiding inappropriate personal services relationships (discussed below) with contractor employees and establishing shared objectives for the teams is the best way to avoid conflict at the personal level. Awareness of other potential areas for concern also is very important.

One area infrequently mentioned as a concern is the loss, not just of core skills, but also of institutional memory and continuity of service. The emergence of the multisector workforce coincident with the ongoing retirement exodus cries for the need for knowledge management practices (chapter 3) to retain knowledge in the government that the government needs to know.

As to other concerns, NAPA has initiated critical work to study the issues surrounding the multisector workforce, with the "immediate focus" to be "an analysis of management challenges relevant to executives and managers working with contractors and subcontractors to the federal government."[8] These challenges have been categorized within six mission-critical areas: accountability, acquisition, human capital and management, social equity and values, legal and governance issues, and organizational culture. Within each area, NAPA posed key questions in a paper released in November 2005.[9] Those key questions that seem most relevant to performance-based project management are listed below (the complete list is in Appendix n.)

Accountability

- Who should be held accountable for the accomplishment of federal missions performed by workers from other sectors?
- How can they be held accountable?

- How do we assign roles and responsibilities to the federal manager to ensure accountability for the performance of the multisector workforce, not just that of federal employees?

Acquisition

- How do we develop, implement and evaluate contracting vehicles to ensure agencies have needed competencies, obtain surge capacity, acquire needed flexibility and resolve specific issues?
- What is the impact of the government no longer directly employing the workforce that is conducting new research and development and building new innovations and technology?
- Are there emerging best practices for acquisition by government agencies that use multisector workforces?

Human Capital and Management

- What special skills are needed to manage a multisector workforce?
- How do we build project management capacity and acquisition skills needed to improve our management of federal contracts?
- How do we sustain core competencies to ensure effective project management, oversight and termination of contracts if necessary?
- How can we help managers avoid pitfalls such as supervising contractors and allowing contractors to provide personal services?
- What is the impact of contractors supervising other contractors on behalf of the federal government or supervising federal employees?

Social Equity and Values

- What impact does contracting out have on the values and goals inherent in the federal government's treatment of its own workforce? If there is a negative impact, can or should it be remedied?
- Does it make a difference that federal employees take the Oath of Office, while contractor employees do not?

Legal and Governance Issues

- How do we stay true to the public purpose of administrative laws such as freedom of information, open meetings, enforcement proceedings, avoidance of conflict of interest and public participation with respect to the activities of government contractors? How do we address this issue across Federal, state and local government lines?

■ What constitutes a "coherent framework of laws," management principles, and organizational practices to assure that government officials have the tools they need to account for the work of the government?

■ How can we determine the best alternatives for establishing a governance structure for the multisector workforce?

Organizational Culture

■ Currently, government managers have limited knowledge of the rules and norms by which the private sector operates. What do managers need to know about the private sector to be effective managers of the multisector workforce? What do managers need to teach the private sector about government?

■ What happens to the culture of an organization when different employees are working under different pay and benefit plans?

■ What measures and performance management systems should be employed with the multisector workforce to ensure effectiveness of that workforce?

NAPA is continuing its study, and answers to some of the questions are beginning to form for the strategic, agency-wide perspective. For example, in its 2006 report, NAPA cites NASA's objectives to (1) establish a workforce planning governance model—with clear roles and responsibilities—that address its total workforce needs (hire and buy), and (2) analyze the sector workforce competencies against needed workforce competencies. In addition, NAPA suggests the "the agency's acquisition planning processes must be integrated with overall strategic and workforce planning, and internal acquisition expertise must be sufficiently available, effective, and efficient" and that NASA "limit vulnerability to personal service relationships."[10]

What are personal services?

Under the Federal Acquisition Regulation (FAR), a personal services contract is characterized by the employer–employee relationship it creates between the government and the contractor's personnel. The government normally is required to obtain its employees by direct hire under competitive appointment or other procedures required by the civil service laws. Obtaining personal services by contract, rather than by direct hire, circumvents those laws unless Congress has specifically authorized acquisition of the services by contract.

In practice, services contracts *not* awarded under the statutory authority for personal services may edge there or end up there as a result of the contract's terms *or the manner of its administration during performance.* The key question is: *Will the government exercise relatively continuous supervision and control over the contractor personnel performing the contract?* Note, however, that giving an order for a specific article or service (deliverable), with the right to reject the finished product or result, is not the type of supervision or control that converts an individual who is an independent contractor (such as a contractor employee) into a government employee.

In its study of NASA's multisector workforce, NAPA advises that federal managers typically can avoid potential problems by:
- Physically separating civil servants from on-site contractors
- Using identification badges to differentiate contractors from civil servants
- Documenting the federal organizational structure clearly
- Specifying the delegation of duties and responsibilities in organizational documents[11]

Additional best practices are suggested later in this chapter because it is our view that the inappropriate use of personal services approaches and relationships under nonpersonal services contracts—and even performance-based contracts!—is what "blends" the multisector workforce and "dulls the line" between the federal and contractor workforce.

Should the rules for personal services contracting be changed?

No, in part because the sectors themselves operate under different laws, authorities, regulations, processes, incentive and disincentive systems, pay and benefit systems, and other factors too numerous to list. The "separateness" of the sectors should be understood, respected, and preserved. It is precisely the "blending" itself and the dulling of the lines that cause problems.

A personal services contract is defined by the FAR as "a contract that, by its express terms or as administered, makes the contractor personnel appear to be, in effect, Government employees." When is that condition desirable? Rarely.

As advocates of performance-based acquisition, we believe that the benefits of the multisector workforce can be tapped while avoiding inappropriate personal services arrangements and, longer term, establishing more strategic workforce management practices. The following chart illustrates some concepts:

Sample Strategies for Taking Advantage of a Multisector Workforce		
Benefit Category	**Benefits**	**Examples of Strategies**
Workforce Management	• Contract out work that is not inherently government • Focus the federal workforce on core missions • Focus contractors on shorter term projects with a limited period of performance • Reduce the size of government • Meet workload demands, sometimes in the face of constraining FTE ceilings or hiring freezes	• Ensure there is a federal workforce, sufficient in numbers and expertise, to manage the contractor workforce—*overall* as well as on a contract-by-contract basis • Elevate the importance of contract management as a crucial federal role, consistent with the fact that so much federal program work is now performed by contractors • Examine the implications of long-term contractor support carefully, with particular regard to core federal expertise, continuity of service, and knowledge management • Examine key workforce indicators, including demographics, turnover, and retention rates, at least annually for each sector—in some agency programs, the stability and retention of the contractor workforce exceeds that of their federal program counterparts • Make sure the federal program is "right sized" and the use of other sectors' workforces is rationale, even in the face of ceilings and freezes • Establish strong program management and program documentation in both the government and contractor sectors • Employ knowledge management practices
Economic	Saving Money	• Examine the economics carefully; it is exceedingly difficult (for many reasons) to calculate the true cost of government performance versus contractor performance
Skills	Acquire knowledge, skills, and abilities absent or in short supply in the current federal workforce	• Examine both the short-term and long-term implications • Determine whether the needed knowledge, skills, and abilities are required long term in the federal government • If they are, use the contractor workforce as a bridge • If not, use contractors on an as-needed basis
Flexibility, Speed, and Mission	• Obtain surge capacity and the ability to scale quickly • Achieve operational flexibility • Adapt quickly to changing mission needs • Augment capacity in the event of an emergency • React to increasing mission complexity • Respond quickly to changing and emerging technology with less direct and negative effect on the federal workforce	• Examine both the short-term and long-term implications • Recognize that contractors generally have fewer impediments to rapid hiring, deployment, redeployment, and termination than do federal programs; corporate success depends on it • Establish multiple long-term contracts with companies whose businesses mirror needed knowledge, skills, and abilities—and compete requirements at the task level on a performance basis • Understand that sizing requirements too small, too short, with too much churn may end up costing the government in terms of the quality and stability of the contractor workforce

Is a time-and-materials contract inherently a personal services contract that contributes to the blending of the workforce?

No. A time-and-materials (T&M) contract is not, in and of itself, a personal services contract. However, the way in which it is structured and managed can turn it into one. Therefore, contracting officers need to be cognizant of the characteristics of personal services contracts and take steps to ensure they don't inappropriately create such a situation.

In general, T&M contracts are the least preferred contract type because the contractor takes few risks and there is no guarantee of a completed deliverable by the time all the hours are used up. In the absence of specified deliverables, federal managers may fall into the trap of managing people, rather than managing work, schedule, and costs. The basic test is whether, in the contract itself or in its administration, you have created an employer–employee relationship, blurring the lines in a multisector workforce.

One of the ways agencies avoid even the appearance of an improper personal services contract is to ensure that only the contracting officer's technical representative or the contracting officer provides direction to the contractor, and that the contract specifically names the contractor representative who will be the sole individual who can accept such direction. No instructions or directions should be given to a contractor employee other than the contractor's named representative.

Thus, T&M contracts are not, by their nature, personal services contracts and can readily be managed in a way that avoids an employee–employer relationship.

Are there any stories of contracts for nonpersonal services that evolved to personal services contracts?

Yes, from a Government Accountability Office (GAO) protest.[12] The situation involved the challenged cancellation of a solicitation for clerical and administrative support services where the government agency's actual requirement was for personal services. Although the contract started off small and included temporary, short-term positions for a limited portion of the agency, the requirements quickly grew into permanent clerical and administrative positions throughout the agency. The contractor's personnel in these positions worked at the agency's offices alongside government employees performing the same or similar work and using government supplies and equipment. Government managers supervised contractor personnel by directing, reviewing, and approving their work. In addition, government managers interviewed and selected contractor personnel for assignment to positions, and routinely requested pay increases and promotions for contractor personnel.

GAO decided cancellation of this solicitation was reasonable and proper because the agency no longer needed temporary personal services for short-term positions; rather, the purpose of this contract was to satisfy the agency's need for full-time, permanent staff. As indicated in this case, when the government, during contract administration, directs or supervises contractor employees over a sustained period, an unauthorized personal services contract may result.

What judgment is applied to determine whether something is a personal service and how might that inform thinking about how to manage in a multisector environment?

The Department of Health and Human Services Project Officers' Contracting Handbook[13] (Exhibit II-I) lists certain factors that address the type of judgment to apply in determining the potential for personal services. The sections below cite some of those factors, with additional thoughts about avoiding the potential for personal services through effective management.

The Nature of the Work

- To what extent does the private sector have specialized knowledge or equipment that might be useful? The solicitation should specify the specific knowledge, skills, abilities or equipment needed by the government, as well as associated deliverables and performance measures and metrics. The government should evaluate the skills necessary to manage contractor performance and identify a federal employee to perform those functions as the contracting officer's representative (COR).
- To what extent does the service represent the discharge of a governmental function that calls for the exercise of personal judgment and discretion on behalf of the government? This factor establishes the service as inherently governmental, and the service cannot be contracted out. *It must be performed by a federal employee.*

■ To what extent is the services requirement continuing rather than short-term or intermittent? Short-term or intermittent services are more likely to be good candidates for contracting out. If the need is continuing, it is important to consider such factors as: (1) maintaining sufficient core functions in house, (2) determining how continuity of service will be achieved, (3) retaining knowledge in house, and (4) deciding how performance of the continuing functions "side by side" will be managed.

Purchasing Provisions Concerning the Contractor's Employees

■ To what extent does the government specify the qualifications of, or reserve the right to approve or disapprove, individual contractor employees? It is permissible to some extent to specify the technical and experience qualifications of these employees, if this is necessary to ensure satisfactory performance. However, it is better to express requirements in terms of performance outcomes and then approve or disapprove performance (deliverables), not people. Reserving the right to approve individual contractor employees is a "red flag" warning of the potential for personal services.

■ To what extent does the government reserve the right to assign tasks to and prepare work schedules for contractor employees during performance of the order? This is another "red flag" warning. It is far preferable to establish work schedules for the contractor in the contract or mutually at the beginning of performance.

■ To what extent does the government retain the right (whether actually exercised or not) to supervise the work of the contractor employees, either directly or indirectly. This is another "red flag." Instead, establish an arrangement whereby supervision of contractor employees is directly managed by a contractor employee. Do not attempt to manage "indirectly" either. This does not, however, preclude the government program manager from sharing observations and feedback—positive and negative—with the contractor manager.

■ To what extent does the government reserve the right to supervise or control the method by which the contractor performs the service, the number of people it will employ, the specific duties of individual employees, and similar details? Care must be taken here. *Sometimes* the method of performing the service must be dictated, such as when the contractor's employees must comply with regulations for the protection of life and property. But it is far better to establish this in the solicitation and contract rather than by verbal direction. Likewise, it is permissible to specify a recommended, or occasionally even a minimum, number of people the contractor must employ, if this is necessary to ensure performance or to help the contractor scope the requirement. However, a far more effective approach is to establish the government's intended results or outcomes and let the contractor propose the "how" and "how many."

- To what extent will the government review performance by each individual contractor employee, as opposed to reviewing a final product on an overall basis after completion of the work? There are few functions that establish a personal service relationship more compellingly than federal employees conducting performance evaluations of contractor employees. Don't go there.

- To what extent does the government retain the right to have contractor employees removed from the job for reasons other than misconduct or security? As long as the reasons are professional and do not violate federal laws (such as those involving equal opportunity or nondiscrimination), this is a reasonable right to retain–and contractors typically are quite responsive.

Contract Performance

- To what extent are contractor employees used interchangeably with government personnel to perform the same functions? This situation may exist, but be mitigated by the manner of supervision and management. It's less a matter of the function performed than of the relationships established.

- To what extent are contractor employees integrated into the government's organizational structure? This is not fatal flaw in and of itself. The FAR statement of guiding principles at FAR 1.102(a) indicates, "Participants in the acquisition process should work together as a team and should be empowered to make decisions within their area of responsibility." That contractors are part of that team is made clear in FAR 1.102(c): "The Acquisition Team consists of all participants in Government acquisition including not only representatives of the technical, supply, and procurement communities but also the customers they serve, and the contractors who provide the products and services."

- To what extent are any of the elements above present in contract performance, regardless of whether they are provided for by the terms of the order? Such elements include supervision or evaluations of contractors by federal employees, approval or disapproval of contractor hiring of employees, routine assignment of tasks to contractors by federal employees, and other activities that signal an employer–employee relationship.

Agencies rely on contractors to perform certain functions in support of their mission goals. For most agencies, contract laws prohibit contracting for personal services. A personal services contract is characterized by an employer–employee relationship. While an agency must judge each requirement, contract, and method of performance on its unique facts and circumstances, the degrees of supervision and control over contractor personnel are key factors to consider when determining if it is a personal services contract.

Should a strategic approach be undertaken to decide when and how to engage the private sector?

Absolutely. In fact, such an approach is essentially required by Office of Management and Budget (OMB) Policy Letter 92-1 which addresses the definition of "inherently governmental" functions that cannot be contracted out. Stan Soloway spoke on this subject to the Acquisition Advisory Panel:

> When I served on the Commercial Activities Panel, we spent a fair amount of time reviewing and discussing the current definition of inherently governmental as contained in OMB Policy Letter 92-1. We came to the conclusion that we could not improve on that policy letter; that the existing guidance was about as prescriptive as one could or should get.
>
> Nonetheless, while changing the regulatory definitions may not be necessary, the current environment clearly demands that we give more consideration to the larger question of roles, responsibilities and accountability.
>
> It requires us to think in terms of the three tiered nature of government work: first, what we might call governance—setting policy, committing government funds, awarding contracts or otherwise legally bind-

ing the government—all of which is clearly inherently governmental; second, areas like contract management and administration, technology assessment, program management, and the like—less clearly inherently governmental functions but clearly areas in which the government must maintain a robust residual capability to ensure that it meets its responsibilities for cost, schedule, performance, and more; and finally, those activities that can be performed either internally or externally.

Of course, this tiering is nothing new. What is different is the degree to which that middle category of government work is tilting increasingly toward contracting. As the government shifts, inexorably and likely irreversibly, to being the manager of service delivery rather than the actual deliverer of those services, this trend will continue.

This is not an innately negative change, although some seem to perceive it as such or characterize it as such for their own purposes. Instead, it simply creates a reality to which we must adjust. It requires us to manage our acquisition and other organic government assets differently than we did in the past and to strategically focus those limited assets where they are most needed. It requires us to recognize the changing nature not only of the supplier base but also of the way in which needed support is delivered to the government and how government can and must optimize its delivery of services to its citizens. We cannot isolate the government from the dynamics of the marketplace. Rather, the government has to adjust to those dynamics.[14]

In its February 2007 report, "NASA: Balancing a Multisector Workforce to Achieve a Healthy Organization," NAPA proposed a decision guide "to help the agency focus on the most important work criteria for deciding whether to use civil servants or contractors."[15] The recommendations are consistent with NASA's determination of seven vital activities to keep in house: strategic guidance, oversight, fundamental decision making, sustaining program momentum, retention of institutional memory, decisions about cost trade-offs, and architectural understanding.

NAPA's Civil Servant-Contractor Decision Guide was designed as a flexible matrix-based tool that includes eight major categories of critical importance: function, resources, workload, labor market, accountability, risk, quality/service level, and employment/flexibility. Within each of these categories is a series of statements, such as:

- The work in question is a critical public function.
- The work is a peripheral competency or function, or a core capacity for which NASA needs temporary augmentation.
- Increased internal competition is desired; for example, to stimulate creativity and innovation.
- Financial incentives may maximize the achievement of an established goal.
- The workload is projected to be stable over time.
- Surge capability and the ability to contract the workforce quickly are required.
- The requisite skills/expertise are available commercially in sufficient quantity/quality.
- There is a possibility of a real or perceived conflict of interest.
- Accountability is project-focused.

The matrix is arranged so that the categories can be weighted and the statements tested against a Likert scale from 1, strongly disagree, to 5, strongly agree. NAPA cites the benefits of using the guide as:

- Consistent understanding of the criteria for deciding whether to hire a civil servant or to buy contractor services
- Transparent decision making and implementation
- Quantifiable results to facilitate decision making, with the opportunity for NASA to weight categories according to agency assessments of relative importance
- Facilitation of meaningful discussions about such as issues as the most appropriate ways to balance the workforce and/or how to address any circumstances restricting the agency's options.

NAPA's Civil Servant-Contractor Decision Guide is available on line at http://www.napawash.org/NASA_Report_2-26-07.pdf.

What skills or competencies are required to manage and lead in a multisector environment?

Noting the 2005 NAPA study, that is exactly what the Centers for Medicare and Medicaid Services (CMS) wanted to know. CMS contracted with Public Sector Communications to study the issue through a combination of interviews, discussion groups, and online research.

In its report, Public Sector Communications noted the change in the role of the federal manager: "Now they must manage—and get to perform—a workforce that doesn't directly work for them and answers to different rules and incentives. To do that requires some new twists on some old skills and competencies."[16]

The study identified ten skills and competencies, then rank ordered them as "most important according to the respondents,"[17] as follows.

Ranking	Skill and Competencies	%
1	Leadership	31%
2	Interpersonal and Relationship Management	19%
3	Oral Communications and Presentations	12%
4	Negotiation and Conflict Resolution	9%
5 - 7 tie	• Program and Project Management • Written Skills • Subject Matter Expertise	6% 6% 6%
8	Policies and Regulations	5%
9 - 10 tie	• Contract Management • Acquisition and Procurement	4% 4%

Again quoted from the Public Sector Communications report are the following explanations of the skills and competencies:

1. Leadership

Leadership is a process by which a person influences others to accomplish an objective and directs the project in a way that makes it more cohesive and coherent. Leaders in a blended workforce carry out this process by applying their leadership attributes, such as beliefs, values, ethics, character, knowledge, and skills.

Although a manager has authority, this *power* does not make a leader, it simply makes a *boss*. Leadership differs in that it makes the staff and contractors *want* to achieve high goals.

2. Interpersonal & Relationship Management

This is the process by which managers build and maintain a positive working rapport with their staff members and contractors to achieve objectives.

3. Oral Communications & Presentations

This is the way managers present to and speak with staff and contractors with the goal of being understood clearly and getting positive reaction and desired results.

4. Negotiation & Conflict Resolution

This is the process of coming to positive agreement on contracts, objectives and scope of work, while at the same time solving contract and personnel issues that arise within a blended workforce.

5. (tie) Program & Project Management

This is everything that is part of the overall process of working with staff and contractors to meet goals. This includes the "soft skills" such as speaking and writing, as well as adhering to policies & regulations, developing acquisition strategies & methods and implementing contract management and performance measures.

6. (tie) Written Skills
This is the process of writing SOWs, emails and reports that are understood by staff and contractors.

7. (tie) Subject Matter Expertise

This is knowing their program inside and out.

8. Policy & Regulations

This is knowing the policies and regulations that impact their program and how they can manage and relate with staff and contractors (e.g., the difference of what they can ask their staff and contractor to do directly).

9. (tie) Contract Management

This is developing and implementing a contract management plan that includes policies, procedures, expectations and outcomes desired from staff and contractors.

10. (tie) Acquisition & Procurement

This is developing and implementing a plan that ensures CMS staff and contractors have the tools and resources needed to meet program objectives.

It is interesting to note that soft skills far outrank domain knowledge on the list, with leadership at the top. Quoting from the report, "Many CMS and government executives said leadership by government managers, whether empowered by title or by actions, is the key ingredient to managing government staff and the relationship with the contractor."

What are the key processes and best practices followed to manage a multisector workforce and optimize performance?

Given that the government must rely on contractor support, for many good reasons, the following are key processes and best practices to manage a multisector workforce and optimize performance.

21 Best Practices for Managing a Multisector Workforce

1. *Establish a governance structure and project plan with clear roles, responsibilities, authority, rules, and reporting requirements.* Be especially sure to address core government responsibilities, such as decision making and acceptance of contractor deliverables.

2. *Establish shared goals and objectives.* While the contractor's goals and objectives are outlined in the contract, a concerted effort may be necessary to align the team. This is especially important when contractor staff must be structured to deliver support that the government cannot. Mission should drive the performance measures of the federal employees, and mission should drive the performance-based outcomes of the contract employees; they should be in alignment. If federal employees' performance goals and measurement standards are different, the team may not perform with unity of purpose. If the

contract is not performance based and no clear performance expectations are established by the contract, the team may not perform with unity of purpose. Also, the government must hold the contractor accountable for the performance objectives of the contract—and ensure that the government's actions do not take the contractor "off the hook."

3. *Be clear about accountability, recognizing that ultimately the federal project or program manager is accountable for overall program results.* Contractors are accountable under the terms of the contract.

4. *Beware of organizational conflicts of interest (OCIs).* Simply stated, an OCI is a situation that occurs when: (1) a contractor is unable, or potentially unable, to render impartial assistance or advice to the contracting agency, or (2) a contractor has an unfair competitive advantage for a contract award. The OCI regulations are designed to help contracting officers recognize and avoid or resolve such conflicts. These regulations focus on four circumstances that tend to give rise to OCIs:

■ *Providing systems engineering and technical direction.* When a contractor provides systems engineering and technical direction for a system but does not overall contractual responsibility for its development, integration, assembly, or production, that contractor is prohibited from competing for a contract to supply that system—as either a prime or subcontractor.

■ *Preparing specifications or work statements.* With certain exceptions, a contractor that prepares or assists with preparing a work statement for a government requirement cannot also compete for the contract award for that requirement.

■ *Providing evaluation services.* Contractors cannot evaluate their own proposals. Nor can they evaluate proposals from their market competitors without proper safeguards to ensure objectivity to protect the government's interest. On this point, it is difficult to imagine what "proper safeguards" would be from the competitors' viewpoints.

■ *Obtaining access to proprietary information.* When a contractor requires proprietary information from other contractors to perform a government contract, and can use the leverage of the contract to obtain it, the contractor may gain an unfair competitive advantage unless restrictions are imposed. These restrictions protect the information and encourage companies to provide it when necessary for contract performance.[18]

5. *Be sure team members understand the standards of conduct for federal employees—and that contractors understand what actions are prohibited at certain levels* (such as picking up a lunch tab). Personal relationships should respect professional boundaries and differences.

6. *Create team space for contractor employees performing on site.* Do not intermix federal and contractor employees whenever possible. This helps differentiate the sectors and reinforces their organizational alliances. Provide contractors badges and physical access needed to perform their contractual obligations.

7. *Differentiate between federal and contractor e-mail addresses for those contractors working on site.*

8. *Ask for information about the contractor's ethics program.* Most, if not all, federal contractors have an ethics program. For example, the Defense Industry Initiative (DII) is a consortium of U.S. defense industry contractors that subscribes to a set of principles for achieving high standards of business ethics and conduct.[19]

9. *Consider including contractors in the agency's annual ethics training.*

10. *Have contractors execute nondisclosure agreements.*

11. *Be sure team members understand the prohibitions on inappropriate personal services.*

12. *Evaluate results—but not contractor employees.* The quality assurance surveillance plan establishes how performance will be evaluated.

13. *Avoid situations that create the appearance of personal services.* Leave contractor staffing decisions to the contractor. Do not ask for resumes or to interview prospective contractor team members. If the level of contract performance is unacceptable, communicate that to the contractor manager of the contract employee. It is in the contractor's interest to satisfy you, the client.

14. *Do not treat contractor employees as federal employees.* It is up to the contractor's management to approve timesheets and leave, evaluate performance, apply discipline, award bonuses, and adjust pay. On the other hand, do not fail to provide feedback on performance, both positive and negative, where warranted.

15. *Recognize that contractor employees, like federal employees, have the right to pursue their own advancement.* The contract doesn't buy that person because it is not a personal services contract.

16. *Do not treat any sector's personnel as "second class" citizens.* Respect the value that multisector perspectives can deliver and quickly handle issues as they arise.

17. *Make plans to capture and transfer knowledge.*

18. *Communicate, communicate, communicate!* It is said that if you want to communicate change, you must communicate it seven times in seven different ways. And be sure you communicate to the entire team. Here is one example: a contractor team was operating in a client setting under a performance-based contract, but the federal employees did not know that. They kept asking "What are they doing?" and tried to assign their work to the contractor's team members, which was out of scope and which the team could not accept. Once the team's federal counterparts understood the contractor's role, the work relationships became much improved. (This example also goes to roles and responsibilities.)

19. *Operate with a "no surprises" standard.* To use a favorite quote, "Bad news rarely ages well." Have a process for elevating problems.

20. *Be careful when you have more than one contractor supporting you as their objectives may not be in line.* In fact, the closer their industries are, the more likely there are to be conflicting objectives. In one situation, an agency had hired two companies—Company A and Company B—in support of a project. Company B was in charge of the project management plan. When Company B missed its deadlines, the schedule was modified quietly. When Company A missed a deadline, there was a lot of finger pointing. One quiet discussion by Company A with the COR fixed that problem. The inequity

was in the unequal treatment by Company B. The objective needed to be "support the client."

21. *And the Golden Rule: Get management buy-in on both sides to this list or something like it.*

SUMMARY

There are many advantages to using contractors as part of the multisector workforce—however, how those contractors are used and managed requires discipline and a strategic approach. The next chapter addresses the need to effectively capture, transfer, and reuse the knowledge of the workforce to improve organization and project performance results.

QUESTIONS TO CONSIDER:

1. How effective has your organization been in avoiding personal services relationships under nonpersonal services contracts?

2. Do you believe new skills are needed to manage effectively a multisector workforce?

3. Have you seen those skills—or witnessed highly skilled performance—in managing the multisector workforce?

4. What other "red flags" on personal services have you observed?

5. Can you think of other best practices?

Endnotes

[1] *Managing Federal Missions with a Multisector Workforce: Leadership for the 21st Century,* National Academy of Public Administration, November 16, 2005, *http://www.napawash.org/about_academy/MultisectorWorkforce12-13-05.pdf.*

[2] *People Policy, & Profit: Driving Value in a Dynamic Market,* 2006 PSC Services Sector Review, Professional Services Council.

[3] *Statement to the Acquisition Advisory Panel, November 18, 2005,* Stan Soloway, President, Professional Services Council, *http://www.acqnet.gov/comp/aap/documents/PSC%20statement%20revised%20-16%20Nov%2005.pdf.*

[4] Ibid, note 2.

[5] Ibid, note 1.

6   *NASA: Balancing a Multisector Workforce to Achieve a Healthy Organization*, A Report by a Panel of the National Academy of Public Administration for the U.S. Congress and the National Aeronautics and Space Administration, February 2007, *http://www.napawash.org/NASA_Report_2-26-07.pdf*.

7   Ibid, note 1.

8   Ibid, note 1.

9   Ibid, note 1.

10   Ibid, note 6.

11   Ibid, note 6.

12   Encore Management, Inc., B-278903.2 (Feb. 12, 1999), *http://archive.gao.gov/legald426p8/161678.pdf*.

13   *http://www.knownet.hhs.gov/acquisition/POHandbookSTD.doc*.

14   Ibid, note 3.

15   Ibid, note 6.

16   *Getting Results with the Blended Workforce: A Research Study on the "Multisector or Blended Workforce" conducted for the Centers for Medicare and Medicaid Services*, July 2006, by Public Sector Communications, *www.PubSector.com*.

17   Ibid.

18   Acquisition Directions™ Advisory, *Organizational Conflicts of Interest*, Acquisition Solutions, January 2002.

19   *http://www.dii.org/*.

APPLYING KNOWLEDGE MANAGEMENT TO IMPROVE PROJECT RESULTS

By William S. Kaplan

INTRODUCTION

There are many definitions of knowledge management (KM). However, in our view, it is critically and centrally about three things: (1) improving performance on an individual, integrated project team, and organizational basis through a consistent and disciplined process for capturing and reusing what you know, from what you do; (2) applying the art and science of making effective practices understood, repeated, improved, and extended; and (3) connecting KM concepts and theories into proven best practices to solve critical business challenges.

HOW SHOULD YOU VIEW KNOWLEDGE IN A PROJECT-DRIVEN ORGANIZATION?

Knowledge comprises *all of the information* in the organization plus *all the experience* and insight. Organizations should leverage and focus their knowledge, in context and in ways that are actionable, meaningful, and relevant, to improve individual, integrated project team, and organizational performance and to deliver value to their employees, shareholders, and customers. This view of KM enables an organization's workforce to make the best decisions and provide the best solutions. Figure 3-1 illustrates this view.

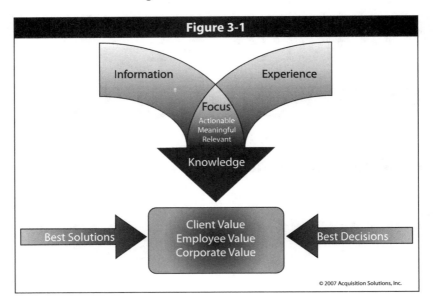

Figure 3-1

Information

Experience

Focus
Actionable
Meaningful
Relevant

Knowledge

Best Solutions

Client Value
Employee Value
Corporate Value

Best Decisions

© 2007 Acquisition Solutions, Inc.

How is knowledge management relevant to performance-based project management?

THREE

Performance-based project management (PBPM) is a knowledge-based process that naturally engenders continuous learning, whether monitoring the transformation of the culture; establishing, maintaining, and sustaining strategic linkage; governing the actions and reactions of all stakeholders; communicating as issues and opportunities arise; managing and mitigating risks as they arise; or managing performance. Thus, there is a natural partnership between the application of the principles and practices of knowledge management and PBPM. This relationship is most effective when the principles and practices of capturing and reusing knowledge are viewed as and implemented as an integrated part of the PBPM process by the organization.

Books and materials about project management contain lots of checklists and suggestions to make your project a success. Some provide people's experiences from their unique situations, and others provide statistical data about projects and outcomes. All of this can be helpful, or not, based on the reader's level of project management experience. For the inexperienced project manager, it can be overwhelming. What is missing?

What often is missing is the advice and insight. What are the crucial success factors? What is the unwritten "know how" and "know why" of the practices? What is the order of things? What do I do first and what do I do next? How are the dots connected? How do I get from A to E on the checklist? Should I focus more on cost than on schedule? And so on....

While an experienced project manager likely will be able to answer some or all of the above questions, a less experienced project manager almost certainly will be challenged. In either case, "knowing what is known" really can help. It is important to make the insight and advice borne from experience accessible and reusable by both the experienced and inexperienced project manager to achieve the organization's mission and objectives. It is also critical to keep the knowledge base of the team as similar and equal as possible, to make sure all are acting on the same information. Knowledge is power when shared. Finally, it's about providing the all important "context" in which the learnings occurred so that relevancy of the knowledge can be determined and the advice and insight can be effectively adapted for use in the current situation.

WHY ARE KM AND PBPM CRITICAL COMPONENTS OF BUYING OR SELLING COMPETENCY?

A great deal of knowledge (information + experience) is used and reused each time an individual or team executes a project. It is important that this knowledge not only is leveraged each time you execute, but also is further developed in a consistent and disciplined manner that captures, adapts, transfers, and reuses what you have learned.

A consistent and disciplined approach to capturing and reusing knowledge helps to develop and improve the competence of the acquisition team in its management and execution of a project. Competence depends on learning not only through training but also through the reuse of experience and insight. Competence, in turn, provides the *agility* for the team to plan for and react to changes in the project while successfully managing overall project performance and achieving broader organization and mission objectives.

Knowledge management should be viewed as a core project management competency. When it is viewed as core, it will be the result of knowledge leadership, because of the actions of the acquisition and project leadership in actively fostering, facilitating, and achieving an organization-wide shared context and understanding of the value of capturing and reusing knowledge as part of the business process. This is critical for sustainable and lasting change and for driving acquisition and project excellence.

WHAT'S THE RELATIONSHIP BETWEEN THE BUSINESS EXCELLENCE VALUE CHAIN AND PBPM?

Knowledge management is an integral part of a holistic approach to performance-based program management. Likewise, PBPM is an integral part of the acquisition management process and continues to be addressed concurrently with project execution. This means that the integrated project team (IPT) must have a systematic framework to capture, transfer, and reuse information, experience, and insight from the first day of the acquisition cycle. A good example of this concept at work is a performance-based acquisition (PBA) that follows the Seven Steps and Six Disciplines of Performance-Based Project Management discussed in this book.

When applied in a consistent and disciplined manner throughout the acquisition cycle, a capture and reuse model, using relevant KM concepts, tools, and techniques, adds value by supporting and helping to establish the foundation for overall acquisition excellence. The outcome is an agile, adaptive, and knowledge-enabled integrated project team (which includes the contractor), successful project management and delivery, and the realization of desired objectives, goals, and results, including improvement to the execution of the PBPM process. Figure 3-2 depicts this concept. Each support activity or component adds value to the outcome, and they span the PBA cycle.

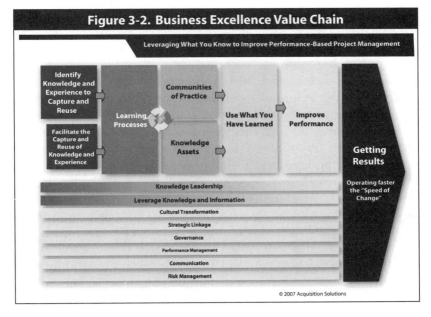

Figure 3-2. Business Excellence Value Chain

Leveraging What You Know to Improve Performance-Based Project Management

Identify Knowledge and Experience to Capture and Reuse

Facilitate the Capture and Reuse of Knowledge and Experience

Learning Processes

Communities of Practice

Knowledge Assets

Use What You Have Learned

Improve Performance

Getting Results

Operating faster the "Speed of Change"

Knowledge Leadership
Leverage Knowledge and Information
Cultural Transformation
Strategic Linkage
Governance
Performance Management
Communication
Risk Management

© 2007 Acquisition Solutions

WHAT DO I NEED TO UNDERSTAND ABOUT KNOWLEDGE MANAGEMENT TO IMPROVE PROJECT EXECUTION?

Helpful to successful PBPM is understanding the strategy we call **Knowledge Convergence.**[1] Knowledge Convergence is the disciplined adoption of a systematic framework to capture, transfer, and reuse information, experience, and insight to measurably improve project performance and customer, employee, and organization value.

When applied to project execution, this strategy enables you to: (1) leverage knowledge in ongoing projects or engagements to immediately improve your IPT's performance, (2) improve your ability to learn from past challenges and successes, and (3) create long-term value from knowledge, experience, and insight held by IPT members.

This strategy and its implementing model rely on several evolved learning points, as follows.

1. Leveraging knowledge is more about people and what they know than technology. Technology and information management alone cannot be relied on for success for two reasons. First, people typically want to just "get their work done," and they may not go through the extra steps required to learn how to use knowledge management technology or tools. Second, there must be a well thought out process for defining, capturing, and reusing the relevant knowledge gained from execution.

2. Knowledge capture and knowledge reuse must work within the context of workflow. When knowledge is captured within the context of an operational or acquisition process as "part of the way people work," it is more meaningful and easier to integrate. When organized and accessible in a way that makes sense to the knowledge users (the team), it then adds value. Technology and tools cannot in and of themselves provide effective "context of use" and add this value.

3. Knowledge capture and knowledge reuse must work within the context of organizational culture. Collaborative cultures provide a better foundation for knowledge convergence than highly competitive cultures. Learning and sharing knowledge in and across all teams must become a routine part of the way an organization works, resulting in open behavior, trust, acceptance of change, and immediate improvements in project and organizational performance due to the agility that this provides.

Sustainable continuous improvement in business and operational processes must be tied to organizational performance measures of success. For example, in the PBPM process, success would be determined based on relevant acquisition process measures, with results easily attributable to applying this model. For example, did the contractor's performance improve in terms of cost, schedule, and quality as a result of the KM practices adopted as an inherent part of performance management? Did questions asked during performance reviews, such as "What obstacles can the government team remove to enable higher efficiency performance by the contractor?" lead to learning that was relevant, meaningful, and most important, *actionable* by the team?

WHAT ARE THE PROJECT BUILDING BLOCKS TO SUPPORT EFFECTIVE CAPTURE AND REUSE OF KNOWLEDGE?

Knowledge Convergence© integrates an ability to "connect, collect, and collaborate" with a discipline of learning before, during, and after process execution. This can provide a professional acquisition or project workforce with the ability to access in real time not only codified knowledge (effective practices, relevant documents, and templates), but also the most current tacit knowledge (experience and insight) that is the "know how" and "know why" of the practice and subject matter areas.

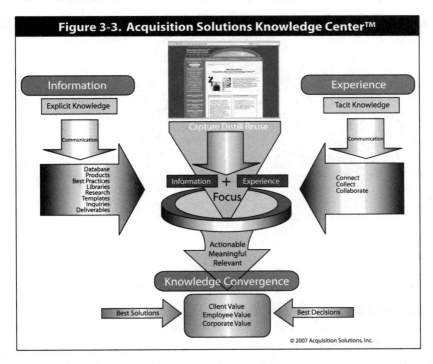

Figure 3-3. Acquisition Solutions Knowledge Center™

© 2007 Acquisition Solutions, Inc.

To enable an acquisition-related or project-based workforce in either government or industry to access this knowledge for reuse, the organization should develop a web-enabled PBPM Knowledge Center.

The PBPM Knowledge Center should be a gateway to an organization's knowledge base, which is accessible 24/7 by its workforce and continually updated with best practices and lessons learned captured from the use of straightforward and effective learning processes, insights captured on video, and peer reviewed knowledge—using the most current workforce knowledge and experience. Figure 3-3 expands on the view of knowledge outlined in Figure 3-1 to reflect a more detailed model of how knowledge should flow and be accessed within a high-performance organization.

CASE STUDY: ACQUISITION SOLUTIONS, INC.

To execute its Knowledge Convergence© strategy, Acquisition Solutions had to create an operating model that balanced four essential elements: learning processes, communities of practice, knowledge assets, and enabling technology (shown as diamonds in Figure 3-4)

Figure 3-4. 4 Diamonds of a Knowledge Enabled Organization

Element 1—Learning Processes: The learning processes enable Acquisition Solutions to capture knowledge and understand it before, during, and after execution or delivery. These processes provide the content for the knowledge assets (knowledge repositories with knowledge artifacts) that reside in the Acquisition Solutions Knowledge Center™.

- **Learning Before (Peer Assists):** "Learning before doing" is supported through the peer assist process, which targets a specific challenge, imports knowledge from people outside the team, identifies possible approaches and new lines of inquiry, and promotes sharing of learning through a facilitated meeting.
- **Learning During (Action Reviews):** A modified version of a U.S. Army technique called after action reviews, this process focuses people on "learning while doing" by having them answer four questions immediately after an activity or event: (1) What was supposed to happen? (2) What actually happened? (3) If different, why are they different? and (4) What can we learn and immediately apply? An added benefit is that, if done well and if people answer honestly, trust builds within the team.
- **Learning After (Retrospects):** "Learning after Doing" is supported by a facilitated process called a retrospect. Conducted immediately after the end of the project or a project phase, a retrospect encourages team members to look back at the project to discover what went well and why, and what could have been done differently, with a view to helping a different team repeat their success and avoid any pitfalls.

Element 2—Communities of Practice (CoPs): Acquisition Solutions has two types of CoPs: organizationally driven and practitioner driven. Both are voluntary and both are encouraged and supported at all levels of leadership. Knowledge from CoPs is harvested and characterized for reuse in knowledge assets so that it can be reused and adapted by fellow practitioners. CoPs help Acquisition Solutions' workforce develop and enhance their competence to contribute individually and collaboratively within their business teams while focusing on the objective at hand.

Success using communities of practice requires broad workforce understanding, the right timing, and support of leadership in terms of time, investment, and resources to participate, maintain, and sustain the communities. Timing is everything, and Acquisition Solutions is always ready because it has socialized and integrated across the company:

- How to view knowledge in an organization (Figure 3-1);
- The framework for how knowledge flows and is accessed for reuse (Figure 3-3);
- The operating model, including tools and techniques, for creating a knowledge-enabled company (Figure 3-4); and

- The desire to embed performing and learning into the fabric of the consulting operation and the greater business and operational processes.

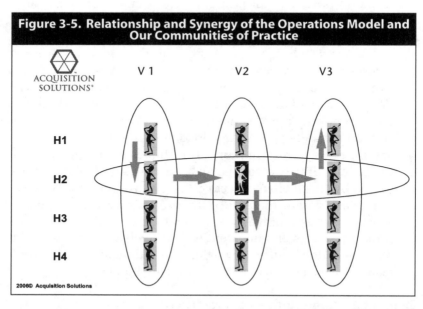

Figure 3-5. Relationship and Synergy of the Operations Model and Our Communities of Practice

2006© Acquisition Solutions

As illustrated in Figure 3-5, there is real synergy between Acquisition Solutions' operations (verticals) and its practice areas and communities of practice (horizontals). Each horizontal practice area has a community of practice that supports the practice and its members. What is learned from execution across the practice or delivery of the solution is available in any vertical or line of business. Conversely, what is learned about delivering the solution and the execution of that practice in any vertical is available across the horizontal and through the community of practice. The intersection of V2 and H2 represents the dual "knowing and performing" transfer capability that this model presents both in concept and operations. It is "connection, collection, and collaboration" at its best.

Element 3—Knowledge Assets: A simple way to describe a knowledge asset might be to call it a repository of knowledge. But it is really more multidimensional than that when constructed and applied effectively. Internally, Acquisition Solutions defines it this way: An Acquisition Solutions knowledge asset is a proprietary, highly organized, and regularly updated compilation of information and knowledge that standardizes practice area methodologies and enables employees to learn and to excel in performance for our clients. Acquisition Solutions' knowledge assets contain:

- Key insights, learnings, and advice in the form of guidelines, checklists, effective processes and practices, and personal stories/short vignettes that highlight critical learnings, insight, or experience;
- The business "context" in which the learning occurred;
- An index of the available knowledge;
- A link to a practice library of documents that can save time;
- Information on finding the person(s) who knows what you need to know when you need to know it; and
- A link directly to relevant training associated with practitioner success in that process.

Element 4—Enabling Technology: Enabling technology is the information technology infrastructure and applications that enable "connection, collection, and collaboration" from any location. The idea is to provide the means for accessing the knowledge and communicating effectively and efficiently with members of the community and others outside the community.

How can understanding the proper application of this model help you to apply knowledge management concepts and practices to improve project and organizational performance?

A few examples of this model at work will provide some insight. We apply the model in many facets of our operations, for example:

Corporate Business Operations

The principles of Knowledge Convergence© are emphasized in all facets of management and execution of operational processes. This means that performing and learning are not limited to buying or selling processes—they apply as well to financial, human relations, corporate development, and other in-house operations. Learning before, during, and after becomes part of business execution.

Delivery of Performance-Based Project Management Solutions to Clients

Prior to key consulting projects, our consulting teams conduct a peer assist to leverage experience and insight from other teams and subject matter experts, enabling the teams to address challenges and risks before they occur. This helps to ensure a more effective and efficient delivery of consulting solutions for the client and a better value for the taxpayer dollars spent.

Improvement to the Performance-Based Acquisition Process

After each project, our consulting teams (often with clients participating) conduct a retrospect to capture for reuse the learnings from the delivery of that project. These learnings are captured and incorporated into knowledge assets so that the changes made become available for immediate reuse by other teams. In addition, the learnings are concurrently incorporated into the company's internal and external training so that the most current knowledge and process efficiencies become immediately available to clients.

WHAT THREE ACTIONS CAN YOU TAKE TO INCREASE THE PROBABILITY OF PROJECT SUCCESS THROUGH KNOWLEDGE MANAGEMENT?

1. **Focus on knowledge leadership.** The foundation for knowledge management is knowledge leadership. Leadership must believe in and support the cultural transformation necessary to enable a performing and learning culture. For any organization, even those that "get it," this is not easy to do, because it requires an investment in time and understanding and a commitment to accept the risk of change. Investment and commitment are required for success.

2. **Recognize the need to "learn as you perform."** Actually deciding to move toward a performing and learning framework that supports results-based project management execution requires executive-level support, an appreciation of the value to be gained, and an understanding of the consequences of not doing anything. Most important, however, it requires individuals and organizations to accept that they do not know everything, that innovation and improvement are always possible, that great learning can come from failure, and that commitment to continual learning greatly enhances the growth opportunities for individuals, teams, and organizations.

3. **Understand that advocacy and accountability are necessary.** As important as top-down support is, execution at the operational level is critical. You also will need advocates and subject matter experts trained in the application and use of

KM concepts, tools, and techniques at the process execution level. They will necessarily perform the following roles:

- Ensure members of a team have access to the knowledge they need to deliver results, wherever this knowledge may have originated;
- Encourage adoption and application of this knowledge in the execution of every project or program in a sustainable manner; and
- Ensure any new knowledge gained through project management is captured and then shared (current and future) in a way that will improve the team's ability to deliver to the organization's goals.

Summary: Applying KM to PBPM

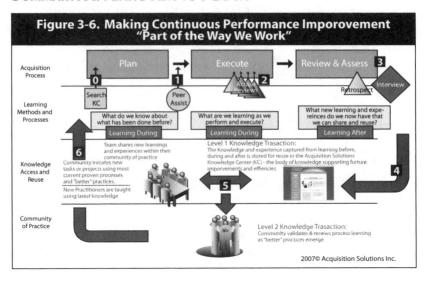

Figure 3-6. Making Continuous Performance Imporovement "Part of the Way We Work"

2007© Acquisition Solutions Inc.

Figure 3-6 provides a road map to put this all together. There are seven steps to consider:

Step 1:

Establish and Understand the Requirements for the Project: This provides the context for your approach to the management of the project. Search for any relevant background you can find within your knowledge resources.

Step 2:

Learning Before (Peer Assist): Understand the planning, integration, and project management support process and timeline. What do you know about what has been done before? Seek knowledge from your peers using the peer assist. Find out what knowledge already exists, where it is, and who has it. Share your project management plan with those peers to gain their insights into challenges you face. Adding their knowledge to yours will create a plan that contains new knowledge (to which you might not otherwise have had access), resulting in new possibilities to consider and increasing the probability of your success.

Step 3:

Learning While Doing (Action Review): Support refinements to the development and professional management of the project and its operational process through your daily performance. This will enable you to address challenges and problems as they occur, making real-time changes for success that can be applied now and in future projects. Ask four questions: (1) What was supposed to happen? (2) What actually happened? (3) If different, why are they different? and (4) What can we learn and do about it today?

Step 4:

Learning After Doing (Retrospect): What happened and why? What did you learn for the future? Complete a retrospect after major project subsections are completed and at the end of the project. The retrospect encourages team members to look back at the project to discover what went well and why—and what didn't and why—with a view to helping a different team repeat their success and avoid any pitfalls. The output of this process provides content for the "knowledge asset."

Step 5:

Create Knowledge Assets: Key project learnings, experience, and effective practices from these learning processes are organized, codified, and packaged in the form of a reusable knowledge asset. The knowledge and experience captured from learning before, during, and after is stored for reuse and accessed by other project managers and teams—the basis for creating future improvements and efficiencies.

THREE

Step 6:

Process Validation and Process Renewal through Communities of Practice: The initial members of this community will be the practitioners involved in the learning processes mentioned above. The project management community validates and renews the process learnings as "better" project management practices emerge. What works again becomes part of the improved and validated process.

Step 7:

Process Continuation: Project managers improve their results on a continuing basis by executing using the most current, proven processes and "better" practices. Measurable results can include:

- Reduced project management and solution delivery costs,
- Reduced project management and solution delivery execution time,
- Improved project management processes and practices that are transferable across the organization,
- Improved overall project management efficiencies through improved access to the right knowledge when needed, and
- Creation of accessible knowledge assets of key learnings, insights, and experience that are readily transferable to evolving project managers and their teams.

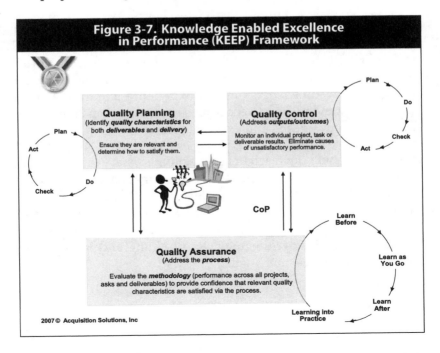

Figure 3-7. Knowledge Enabled Excellence in Performance (KEEP) Framework

Two additional and critical factors to understand are the relation-ship between quality and KM and how this relationship affects project and organizational performance (outcomes and outputs). The quality framework and knowledge management framework should be inextricably linked from the planning of the project, creation of the project management plan, and execution and delivery through the Knowledge Enabled Excellence in Perfor-mance (KEEP) framework. KEEP has three aspects aligned with the knowledge capture and reuse framework: quality planning, quality control, and quality assurance.

- Quality planning refers to identifying the quality characteristics for both *deliverables* and *delivery* and ensuring they are relevant to the project while determining how to satisfy them in the context of the specific client and project.

- Quality control addresses results, both in terms of *outcomes*, what we want or expect to happen as a result of the execution of the project, and in terms of *outputs*, what we will deliver as a result of project management and execution.

- Quality assurance addresses the *process*. We continuously evaluate our project management methodology to provide the confidence that relevant quality characteristics are satisfied through the all aspects of project execution and that outcomes and outputs support the client's requirements and mission.

This model for knowledge capture and reuse has a direct and intentional relationship to the "Plan-Do-Check-Act" quality cycle and the KEEP framework. Our team members "learn before," "learn during," and "learn after," to be able to continuously adapt and apply new knowledge, the new experience and insight, from methodology and process execution, to ensure they can consistently deliver to client's needs and expectations.

Finally, while you can teach the core skills necessary to understand the concepts of performance-based project management, practice or execution competency is experience based. KEEP supports our ability to "operationalize training" by ensuring that the most cur-rent operational experience and insight, which is always in people's heads, is immediately integrated into training and immediately becomes part of practice.

CONCLUSION

To effectively develop and implement a successful knowledge capture, transfer, and reuse framework, the following factors are vital:

- There must be senior leadership advocacy.
- Experienced change agents must be available.
- Delivering specific, tangible, process-driven performance improvements must be a goal.
- You must be prepared to invest in maintaining and sustaining the knowledge framework and model.
- You must embed and integrate a common, simple set of core capture, transfer, and reuse practices that are easily understood, supported, and performed "on the job" and "in the job," to embed a sustainable way of working.
- You should leverage any existing investment in your organization's technology base before looking for outside technology or applications.
- To consistently deliver quality outcomes and outputs, execution and training must be aligned through a consistent and disciplined process that leverages operational experience and insight.

With the right leadership, strategy, and help, it does not have to take a long time or a lot of money to show improved results.

QUESTIONS TO CONSIDER

1. How effectively does your organization capture and reuse knowledge to improve project results?

2. What knowledge management processes, tools, or techniques does your organization currently use?

3. What is the annual cost to your organization of knowledge lost?

4. How can you ensure that what you are learning about project management is consistently and effectively integrated into future project execution?

5. Does your team or organization have what it takes to accept that it does not know everything, that innovation and improvement

are always possible, that great learning can come from failure, and that commitment to continual learning greatly enhances growth opportunities

Endnotes

[1] ©2007 Acquisition Solutions, Inc.

CHAPTER 4

FIRST DISCIPLINE – CULTURAL TRANSFORMATION

By Ann Costello

INTRODUCTION

All six disciplines are important to performance-based project management (PBPM), but cultural transformation may well be the primary ingredient for success because it creates human readiness for change. We use the equation

$$B = f(p, e)$$

to reflect that Behavior is a Function of the Person in his or her Environment. Cultural transformation can shift both the behaviors of people and the environment to a more productive, results-oriented state of being, necessary for results-oriented project management. Shared goals and values promote a results-based business environment that motivates individuals, teams, and organizations to execute their authority and fulfill their responsibilities.

What is culture, and what is the cultural transformation process?

In her article, "The Key to Cultural Transformation," Frances Hesselbein (editor-in-chief of *Leader to Leader*, chairman of the Drucker Foundation, and former chief executive of Girl Scouts of the USA) says, "Peel away the shell of an organization and there lives a culture—a set of values, practices, and traditions that define who we are as a group. In great organizations the competence, commitment, innovation, and respect with which people carry out their work are unmistakable to any observer—and a way of living to its members. In lesser organizations, distrust and dysfunction are equally pervasive."[1]

Another definition of culture provides this view: "An organization's culture encompasses the values and behaviors that characterize its work environment, and in particular, how people work with each other, how they are held accountable, how they are rewarded, as well as how communication flows through the organization."[2]

Cultural transformation has been defined as "the introspective and inclusive process by which an organization formulates its values and revisits its mission [to] allow organizations to serve their customers and communities, with high performance, to be viable and relevant in an uncertain future. That capacity to change and to serve is the essence of a great and vibrant culture."[3]

FOUR

Or as described by Acquisition Solutions, "Cultural transformation [is a process that] encompasses the identification of how an organization and its people need to change to make an initiative successful, as well as the shepherding of that organization through acceptance and institutionalization of the new processes and behaviors."[4]

What is the importance of cultural transformation to getting results through PBPM?

Inherent in the nature of project management is change. Inherent in the nature of people is a varying level of tolerance for change. Inherent in the nature of teams is the progression from forming and storming to norming and performing required for effective teamwork. As reported by the Government Accountability Office (GAO):

> At the center of any serious change management initiative are the people. Thus, the key to a successful ... transformation is to recognize the "people" element and implement strategies to help individuals maximize their full potential in the new organization, while simultaneously managing the risk of reduced productivity and effectiveness that often occurs as a result of the changes.[5]

Cultural transformation readies people, individually and organizationally, for effective and efficient performance. This is especially important for project management teams, which often bring together people from different cultures. Whether it is a program management office for a billion dollar multiagency acquisition project or a project management team comprised from a multisector (blended) workforce, the culture and personal motivation of the team members may vary considerably. So an important readiness step is to build a culture unique to the team and, possibly, for the entire program management office.

This chapter, like the chapters that follow, addresses the key inputs, tools and techniques, and desired outputs of each discipline.

Figre 4-1. PBPM First Discipline: Cultural Transformation
Cultural Transformation Process

Key Inputs	Tools & Techniques	Outputs
Leadership	Select a facilitator	Collaboration vs. Direction
Fairness & Trust	Understand what the culture is today	Insight vs. Oversight
Clarity	Agree on where you want the organization to be	Baselining & Measureing vs. Control
Consistency		Results vs. Compliance
Reinforcement	Document	
	Communicate and reinforce	Partnership vs. Dictatorship
	Invest	Leading vs. Directorship
	Remove obstacles	
	Create ownership & rewards	Facilitating vs. Obstructing
	Live it	Managing the Relationship vs. Paper
	Build leaders	

What are the key inputs of a cultural transformation?

Certain factors are critical to a successful cultural transformation. These include:

■ **Leadership** – There must be powerful believers and motivators in charge. Change requires champions. To steer and drive the project in the right direction, project leads and managers must mobilize staff, manage resources, engage stakeholders, supervise work operations, and oversee management processes. Leaders help the team create shared ownership of the mission, vision, and values, monitor adherence, and also set performance expectations and performance goals aligned with the mission. Leaders need followers: the leader identified as executive sponsor of the project must be someone who is respected by the organization faced with change. GAO offered this perspective on leadership during transformation:

> Because a merger or transformation entails fundamental and often radical change, strong and inspirational leadership is indispensable. Top leadership (in the federal context, the department Secretary, Deputy Secretary, and other high-level political appointees) that is clearly and personally involved in the merger or transformation represents stability and provides

FOUR

an identifiable source for employees to rally around during tumultuous times. Leadership must set the direction, pace, and tone for the transformation.[6]

- **Fairness and Trust** – These are cornerstones of an open environment focused on project success. Organizational leadership should promote, display, and expect "fair and reasonable" behavior, and this behavior should be guarded by the team. Inequity should be avoided because it will destroy trust, organizational credibility, and performance outcomes. Without trust, good working relationships will not form and the sense of teamwork and shared goals will not develop, making successful results-oriented project management unlikely.

- **Clarity** – For individuals or groups to execute their authority and fulfill their responsibilities, they must have a clear picture of what they are to execute and what results are expected. Key areas demanding clarity are authority, organizational mission, individual roles and responsibilities, performance expectations, performance monitoring, definitions of success, and, critical but seldom discussed, the attitude of the team. Attitude is contagious. Optimism and energy breed optimism and energy, especially when displayed by leaders.

- **Consistency** – This ensures stability. An inconsistent application of policies, procedures, resources, and/or consequences within an organization undermines a results-oriented environment by weakening individuals' perceptions of organizational commitment and credibility. It deflates morale and promotes cynicism within a project or organization.

- **Reinforcement** – Leaders and team members should continue sending and sounding the transformation message. Resistance to change and the power of habits are the opponents of transformation. Because the old and familiar way of doing business often exerts a stronger pull than the new, people can backslide into old routines and processes, sometimes without even being aware that they are doing so.[7]

What are the key tools and techniques used to achieve world-class cultural transformation?

Both the need for cultural change and the difficulty in achieving and sustaining it often are underestimated—and efforts to facilitate it frequently are underfunded. You must be prepared to invest if you want to create and maintain the right environment, one that promotes and encourages measuring performance and delivering results; motivates individuals, teams, and the entire organization to execute their authority and fulfill their responsibilities; stimulates people to perform work and achieve the desired results; inspires shared results; and ensures repeatable processes.

The following are ten important steps to achieving cultural transformation:

1. **Select a facilitator and recorder.** A best (and probably essential) practice is to start the cultural transformation using a trained facilitator and recorder who are not stakeholders in the process. Working within the organization, through group meetings as well as one-on-one interactions with staff, these experts can help the team create a road map and accelerate the transformation progress, adding significant value by keeping the effort on track and saving time and effort.

2. **Understand what the culture is.** If the team members come from an organization with a strong, positive culture and a shared sense of values, this step may seem simple. However, even a strong, existing culture should be tested against the mission the team is about to undertake. If the project the team is going to manage will introduce significant change into the organization, then the culture will be tested by the stress of change and the potential backlash from others in the organization. Time should be spent affirming the culture and the team's project objective.

   It is more likely, however, that the team members will come from different cultures, weak or negative cultures, and maybe even different value systems. This can occur, for example, in blended workforce teams with both public-sector and private-sector members. In such cases, creating shared objectives and a team value system is essential, especially if the project contract does not include good incentive strategies. The dynamics of team and human interaction—forming, storming, norming, performing—sometimes create a team culture, but that's leaving too much to chance.

Although widespread organizational transformation may require in-depth assessments, comprehensive surveys and interviews, organizational readiness studies, environmental scans, employee focus groups, gap analysis, and formal training and change management plans, most teams can follow a more simple process: mission, vision, and value exercises. We followed this process when Acquisition Solutions had about 30 employees. While the company founders and "pioneers" lived a culture, it had to be captured and affirmed, or it would have been possible to lose it as the company grew.

3. **Agree on where you want the organization to be.** For example, at Acquisition Solutions, the leadership team met, discussed, and came to consensus on the company's mission, vision, and values. Going through the experience as a team was an important aspect of developing cultural soundness and shared values. Our mission focuses on what we seek to achieve every day, and our vision is our stretch goal that pushes us to continually seek to transform for the benefit of our clients, our company, and our individual personal development. The vision is the "end state," and clarity about what constitutes a successful outcome is important for every team. We ask ourselves in the course of our work, "How will we know we were successful when this project is done?"

The values exercise also was a shared experience. The leadership team sat together and "went around the room," offering up values we thought important to us as individuals and as a company. We created a long list that included such words as optimism, honesty, creativity, excellence, flexibility, integrity, loyalty, respect, results, teamwork, and trust. When we reconvened several weeks later, the task was to decide which three of these many wonderful values we most wanted to adopt and that we believed best captured our essence. The values we chose—integrity, teamwork, and excellence—guide our daily approach to our work and continue to inform important corporate decisions.

4. **Document it.** Every new hire at Acquisition Solutions is given a laminated card, about the size of an index card, with the heading "Strategic Drivers." It includes the company's vision, mission, values, motto, inspiration statement, and guiding principles. It is not unusual to see employees pull the card out of a pocket or purse when talking to each other or to clients. At one of our all-hands meetings, a new employee shared that when his former boss asked why he was leaving to join us, he pulled out the laminated card and said, "Here's why." A positive culture that is shared by the people in an organization creates energy, optimism, and results.

5. **Communicate it and reinforce it.** The rules are simple. First, communicate early, communicate clearly, and communicate often. Second, communicate "the few powerful, compelling messages that mobilize people around mission, goals, and values."[8] Third, when action or behavior is inconsistent with mission, vision, or values, correct it. Remember that your participants will range from early adopters to those who prefer a "wait and see" approach.

6. **Invest in it.** Making the transition to a culturally positive results-based project management team requires an investment in the workforce. For some, the needed leadership and interpersonal skills may not have been acquired or rewarded in the past and may not come naturally or easily. Among the techniques that can be used to develop these skills are one-on-one and group meetings, individual coaching and mentoring, facilitated sessions, and team-building exercises. Formal training may be required, and (depending on scale) so could completion of a skills assessment and the development of individual and group training plans. The more "primed" the team is to function in this results-based environment, the greater the benefit derived and the faster to the finish line. Also, whenever a new member is assigned to the team, he or she should be "socialized" to the culture and trained, to ensure a successful transition into the integrated team.

Team training helps to baseline a common understanding and to build the team. At Acquisition Solutions, when we begin a "Seven Steps"[9] performance-based acquisition project with a client, we conduct a several day orientation session that includes both our employees and federal employees. This approach and the professional development processes mentioned above help to develop a high-performing team that is adaptable and prepared to address current and future challenges.

7. **Remove obstacles.** To build flexible, nimble project management teams "that unleash the energies and spirits of our people,"[10] seek to challenge entrenched processes, organizational impediments, and habits. Don't accept the "this is the way we've always done it" argument. Adopt only those policies, practices, and procedures that conform to mission, vision, and values. Embrace change.

8. **Create ownership and rewards.** Creating a sense of ownership within the team, coupled with authority and a reward system, can give members a vested, personal interest in the outcomes and lead to optimal team performance. GAO has reported that "when people freely share and are rewarded for what they know, they are more likely to feel a stronger connection to the new organization."[11] Ownership creates an interest in project outcomes and leads team members to fulfill their responsibilities. Ownership increases responsible behavior and a caring attitude. GAO reports:

> A successful merger and transformation must involve employees and their representatives from the beginning to gain their ownership for the changes that are occurring in the organization. Employee involvement strengthens the transformation process by including frontline perspectives and experiences. Further, employee involvement helps to create the opportunity to establish new networks and break down existing organizational silos, increase employees' understanding and acceptance of organizational goals and objectives, and gain ownership for new policies and procedures.[12]

The right reward system is important as well. GAO has identified the following key practices for effective performance management:

Align individual performance expectations with organizational goals. An explicit alignment of daily activities with broader results helps individuals see the connection between their daily activities and organizational goals.

Connect performance expectations to cross-cutting goals. Placing greater emphasis on collaboration, interaction, and teamwork across organizational boundaries helps strengthen accountability for results.

Provide and routinely use performance information to track organizational priorities. Individuals use performance information to manage during the year, identify performance gaps, and pinpoint improvement opportunities.

Require follow-up actions to address organizational priorities. By requiring and tracking such follow-up actions on performance gaps, these organizations underscore the importance of holding individuals accountable for making progress on their priorities.

Use competencies to provide a fuller assessment of performance. Competencies, which define the skills and supporting behaviors that individuals need to effectively contribute to organizational results.

Link pay to individual and organizational performance. Pay, incentive, and reward systems that link employee knowledge, skills, and contributions to organizational results are based on valid, reliable, and transparent performance management systems with adequate safeguards.

Make meaningful distinctions in performance. Effective performance management systems strive to provide candid and constructive feedback and the necessary information and documentation to reward top performers and deal with poor performers.

Involve employees and stakeholders to gain ownership of performance management systems. Early and direct involvement helps employees' and stakeholders' understanding and ownership of the system and belief in its fairness.

FOUR

Maintain continuity during transitions. Because cultural transformations take time, performance management systems reinforce accountability for change management and other organizational goals.[13]

9. **Live it.** More important than communicating the mission, vision, and values is living them. If the personal and business decisions and actions of the leadership support the cultural ideals, then those ideals can permeate the organization. Employees at all levels are important purveyors of the corporate culture. Acquisition Solutions once had a business opportunity for a substantial piece of sensitive assessment work that was being negotiated on a sole-source basis. However, when the client negotiated for the right to review and approve the findings of the "independent assessment," we withdrew. We felt we could not comply with the request and remain true to our value of integrity.

10. **Build leaders throughout the team or organization.** Frances Hesselbein recommends "[d]*ispersing* the responsibilities of leadership across the organization, so that we have not one leader but many leaders at every level of the enterprise." By valuing and seeking leadership qualities in each individual, powerful teams and organizations are built.

What desired outputs or standards of behavior are important in managing for results?

Performance-based project management requires an organization and its people to take a different approach to doing business and interacting. It requires:

- **Collaboration versus Direction.** Team members must collaborate for mutually beneficial results. If the project involves the use of contractors, heavy-handed, directed performance removes contractor flexibility and is compliance oriented rather than results oriented. Worse, directed performance can have the effect of shifting the responsibility for performance from the contractor to the government team.

- **Insight versus Oversight.** Reviewing processes and methodologies is not nearly as important as understanding the effect of performance on the desired results. How work is done is not as important as the results achieved. People have different styles of learning and working, but as long as everyone is well informed about the objectives and the state of performance, less oversight is required and true teamwork can develop. At Acquisition Solutions, we have a saying: "Good people with good information make good decisions."

- **Baselining and Measuring versus Control**. Results-oriented project management requires performance monitoring, not process control. Team members must develop the capability and capacity to manage to results.

- **Results versus Compliance Orientation**. In a results-based environment, the focus needs to be on managing objectives, defining success, and measuring outcomes.

- **Partnership versus Dictatorship**. Results-oriented project management requires that all team members act with authority, employing their full potential. Effective project teams have little hierarchy.

- **Leading versus Directing**: Leadership enables the team to implement its vision, applying the resources necessary to achieve the mission. Effective project teams have leaders, not bosses.

- **Facilitating versus Obstructing**: Team members who work together to resolve problems with processes and/or people help to keep everyone's focus on the mission. Don't play the "blame game."

- **Managing the Relationship versus Managing the Paper**: People—not paper—get things done. With regard to managing contracts, for example, at Acquisition Solutions we say, "A good contract is one you put in the drawer and leave there." People and performance take over. With the right alignment and incentives, the success of the government and the contractor are one and the same.

CASE STUDY – GENERAL ELECTRIC

For more than 25 years the values of chief executive officer Jack Welch formed the culture of General Electric (GE). The GE way of doing business was crystal clear: continuously improve operations using Six-Sigma as a tool, always make the performance targets, reduce costs whenever possible, and achieve number one or number two position in your industry. GE was highly successful for decades using this approach.

However, an individual, team, or company's greatest strength can become its greatest weakness. When Jeff Immelt became the new head of GE in 2001 he realized GE's tremendous focus on performance results was limiting innovation. Thus, he recognized that he needed to lead a major cultural transformation to inspire more risk taking, innovation, and marketing, to enable GE to continue to grow and achieve future business success.

To facilitate the transformation, Immelt embarked on a multimedia communications campaign to inform all of GE's employees, suppliers, and customers of the company's new focus on innovation. Next, he changed the performance measures for GE's top leaders to include funding new ideas, leading teams to develop better ideas, and inspiring employees to take risks to achieve high performance.

Stories from the Front Lines

Some of the most important performance-based project management jobs done in the federal government revolve around managing contractor performance to achieve mission results. Two stories from the front lines of federal acquisition can help illustrate.

Performance-Based Acquisition: Boon or Bust?

At Acquisition Solutions' quarterly client forum in November 2006, a panel of experienced acquisition professionals explored the topic: Is performance-based acquisition a boon or a bust? With notable interest and involvement from the audience, the overall conclusion drawn was that performance-based acquisition is a "boon" *when done properly.*

Training needs and cultural issues topped the list of potential inhibitors to conducting successful performance-based acquisitions, to which the audience added the challenges of leadership and program management support. "Senior leadership must buy in, show support, and get the team to do what they need to do," noted one of the panelists. The consensus was that a team-based approach to training is needed, as well as a cultural transformation that brings those involved in the acquisition into the process from the start. Communication plans also were recognized as an integral part of keeping all stakeholders invested. "You need an ongoing campaign, because people have a tendency to go back to what it is they are comfortable with and what they know," stressed another panel member.

What do federal acquisition thought leaders think about the importance of culture?

In October 2006, Acquisition Solutions assembled a "learning after" forum of very experienced federal acquisition thought leaders and practitioners for a more specialized focus: applying performance-based techniques to the acquisition and management of complex systems development. As reported in our November 2006 Acquisition Directions *Advisory*, "Best Practices and Lessons Learned from the Front Lines: The Art of Successful Performance-Based Systems Acquisition," culture emerged as a very special consideration in PBPM:

Recognize the culture issues.

Culture matters. One participant said, "The culture issues are just incredible here. [We are] dealing with strong preferences for certain types of devices, configured in particular ways. We're seeking a new way of doing business . . . without interfering with the mission. And we're taking a performance-based approach."

Another said, "We are fortunate in that we don't have a lot of capital assets." The implication was that there was not an inherent preference for a solution. Further, the acquisition had built in a fast roll test deliverable. "We have a focused effort out of the gate in eight months to do [a phased part of the project]. That will give us an opportunity within eight months to see how this is going to go." He also observed, "We don't have to unlearn; we can grow our own culture from scratch. There is not an entrenched specifications-trained culture."

Another forum participant saw culture as a positive. "We are leveraging all that is good in our culture in law enforcement and taking it to the next step, to connect the dots." His observation of the potential of "evolving culture" recognizes the critical importance of integrated teams in conducting major systems development and implementation strategies.[14]

In the next chapter, we discuss the Second Discipline of Performance-Based Project Management—the practice of strategic linkage.

QUESTIONS TO CONSIDER:

1. How effective has your organization's leadership been in driving your key projects to high-performance results?

2. Would you describe your current organization as consistent, stable, and high performing? If not, why not?

3. How effective is your organization's leadership in removing obstacles that impede success?

4. Has your organization effectively linked pay to individual, team, and organizational performance goals?

5. What organization do you think of when you discuss high-performing organizations?

Endnotes

[1] Frances Hesselbein, "The Key to Cultural Transformation," available on line at *http://www.leadertoleader.org/knowledgecenter/L2L/spring99/fh.html*.

[2] Jeffrey A. Schmidt, ed., *Making Mergers Work: The Strategic Importance of People* (Alexandria, Va.: Towers, Perrin, Foster and Crosby/Society for Human Resource Management, 2002), as reported by GAO in GAO-03-669.

[3] Ibid.

[4] Acquisition Directions™ *Advisory*, "Performance-Based Acquisition Requires the Six Disciplines of Performance-Based Management," Acquisition Solutions, May 2004.

[5] GAO, "Results-Oriented Cultures: Implementation Steps to Assist Mergers and Organizational Transformations," GAO-03-669, July 2003.

[6] Ibid.

[7] Ibid, n. 4.

[8] Ibid, n. 1.

9   *http://acquisition.gov/comp/seven_steps/index.html.*

10   Ibid, n. 1.

11   Ibid, n. 5.

12   Ibid, n. 5.

13   Ibid, n.5.

14   Acquisition Directions *Advisory*, "Best Practices and Lessons Learned from the
     Front Lines: The Art of Successful Performance-Based Systems Acquisition,"
     Acquisition Solutions, November 2006.

SECOND DISCIPLINE – STRATEGIC LINKAGE

By John A. Gaeta

INTRODUCTION

Leading-edge organizations, whether public or private, link their strategic goals to their performance evaluation systems to gain insight into, and make judgments about, the effectiveness and efficiency of their programs, projects, processes, contractors/suppliers, and people. These best-in-class organizations decide what indicators they will use to measure their progress in meeting strategic goals and objectives on a contract, project, program, and portfolio basis. Then, they gather and analyze performance data and use that data to drive improvements, successfully translating data to information to strategy to action.

This strategic linkage provides a consistent vision that cascades throughout an organization and ensures that the results achieved reflect agency/corporate and organizational strategic goals, as well as the objectives of a particular contract, project, or program. In other words, linkage or alignment is having a common agreement about goals and means, both organizationally and personally.[1]

Soichiro Honda, founder of the Honda Motor Company, posited three "Sacred Obligations of Senior Leadership," which he described this way:

- **Vision:** What will we be?
- **Goals:** What four or five key things must we do to get there?
- **Strategic Linkage:** Take action to translate the work of each person into an alignment with the goals.

What is the strategic linkage process?

To maximize performance, best-in-class organizations in both government and industry have recognized the value of linking their strategic organizational goals and objectives to their performance evaluation systems, thus aligning programs, projects, contracts, integrated project teams, and individuals. Figure 5-1 illustrates the strategic linkage process used by best-in-class organizations, including the key inputs, proven tools and techniques, and desired outputs. This chapter provides a brief discussion of each of the key items contained within the strategic linkage process, along with numerous best practices and case studies of successful organizations from both U.S. government agencies and industry.

Second Discipline – Strategic Linkage

Figure 5-1. Performance-Based Project Management

Key Inputs	Tools & Techniques	Outputs
Organizational Framework of Goals and Objectives	Map the Organization's Goals and Objectives	Turn Data into Valuable Information
Organizational Performance Framework	Map the Organizational Performance Framework	Share Information with Employees, Customers, and Stakeholders
Individual Performance Evaluation Framework	Understanding the Individual Performance Evaluation Framework	Use Information to Drive Improvements
Contract Performance Evaluation Framework	Assess the Contract Evaluation Framework	
	Overlay the Maps, Identify the Gaps, and Create the Master Linkage Strategy	
	Maintain a Culture of Performance with Empowerment and Accountablility	
	Develop at All Levels Well-Defined Roles and Responsibilities	
	Create a Quality Assureance and Surveillance Plan (QASP)	
	Align Performance Management for Strategic Linkage	
	Implement Recognition and Reward Programs	
	Implement Performance Measurement Best Practices	

Why is strategic linkage vital?

When strategic linkage is strong, people feel a clear and shared sense of mission, energy runs high, and individual, team, and organizational effectiveness increases. Strategic linkage is particularly evident, for example, at start-up companies where everyone is focused on a few critical business goals. These firms are typified by a deep sense of dedication and intense personal engagement on the part of nearly all involved.

When strategic linkage within an organization is weak, people can end up working at cross purposes, and actions become less effective.

Often, functional or individual objectives take precedence over the needs of the larger organization, mission, or customer. Morale and productivity diminish over time, and the organization fails to achieve its objectives and becomes more vulnerable to competitors and market and sector forces.

On the largest scale and at the highest level of performance, strategic linkage is all about achieving absolute alignment of goals so that all parts and functions of an organization's value chain work toward the same purpose. In its ideal form, strategic linkage allows all members of the organization to align their personal values and objectives with those of the firm.[2]

How do you implement strategic linkage in the U.S. government?

Let's start with how *not* to start—not with the contract action itself. What does that mean?

In the traditional contract acquisition and management environment, we sometimes focus too narrowly on the requisition-generated task at hand, at the expense of the bigger picture. Instead, take a step back to ensure that what you set out to acquire—and how you set out to acquire it—is aligned with your organization's strategic goals and with the promises made in the program's "approval document." The point is that, unlike rote exercises starting with a statement of work, you begin with an understanding of the administration's management mandates and your agency's strategic plan, program approval document, and/or the project's Exhibit 300.[3]

Thus, during the initial stages of a project, to establish strategic linkage, you must link objectives and definitions of success to the organization's mission, vision, critical success factors, and objectives. As illustrated in Figure 5-2, it also is important for federal agencies to link project or program objectives with the agenda of the executive branch of the government—for this administration, the President's Management Agenda.

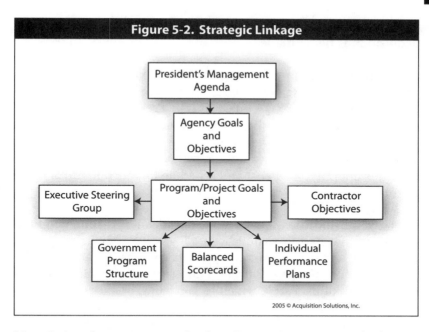

Figure 5-2. Strategic Linkage

2005 © Acquisition Solutions, Inc.

Now let's take it one step further. In conformance with the administration's management mandates and with the Government Performance and Results Act of 1993 (GPRA), each agency must establish performance goals, a five-year strategic plan, and an annual performance plan. GPRA further requires agencies to report annually to the President and the Congress via their "performance-based budget" on program performance for the previous fiscal year, setting forth performance indicators, actual program performance, and a comparison with plan goals for that fiscal year. The purpose is to improve federal program effectiveness and public accountability by promoting a new focus on results, service quality, and customer satisfaction.

The strategic plan is an essential resource document for developing contract objectives, along with the program business case (sometimes an Exhibit 300 prepared for the Office of Management and Budget [OMB]) or a program management plan. Using these documents should ensure proper strategic linkage among the project, the contract, and the mission goals of an agency.

Acquisitions take place within this framework and should align with the goals and the plan. Thus, the investment represented by the acquisition should discuss the agency's mission and strategic goals and the performance objectives sought—in short, why the acquisition is being made.

Done correctly, contract objectives should link to program goals, which should link to the agency's strategic plan, which should link to the agency's GPRA goals and the President's Management Agenda or similar executive branch guidance. In addition, OMB guidelines instruct agencies to combine reporting GPRA results with a business case for funding, which further mandates that agencies not only meet cost and schedule targets but also achieve agency goals.

Every acquisition should be considered an investment. OMB Circular A-11, part 7, establishes policy for planning, budgeting, acquisition, and management of federal capital assets, and provides instructions on budget justification and reporting requirements for major acquisitions. Part 7, which requires the submission of an Exhibit 300 for major or mission-critical capital investments, is designed to collect information that will assist OMB during budget review and serves as a primary means for justifying investment proposals.

Whether or not a project reaches the major program or Exhibit 300 threshold, applying some degree of capital investment scrutiny to the management strategy is a good practice.

Strategic linkage will ensure a focused approach that aligns all tasks toward successfully meeting goals that support the organization's ultimate mission, purpose, and objectives. As changes occur over time, organizational goals and objectives also may change. Thus, it is important, as the project progresses or matures, to ensure that the linkage to the project, program, and organizational objectives is maintained. Continual review is an imperative.

What are the key inputs needed to establish strategic linkage?

To establish strategic linkage at the project level, you will need to understand four things:

- Your organization's framework of goals of objectives
- Your organization's performance framework
- Your organization's approach and framework for individual performance evaluation
- The contract's approach to performance goals, metrics, and evaluation

FIVE

1. Organizational Framework of Goals and Objectives

You need to locate all the documents that together create the organizational framework of goals, objectives, and plans for performance. What do the President's Management Agenda and other administration documents say that relates to the project you are leading? What do your agency's strategic and annual performance plans say and how do they relate? Has your agency released statements to the press or public that touch your program or project? How do department-level or agency-level goals relate to those of your organization or program? Are new goals introduced at the program or project level?

2. Organizational Performance Framework

You also need to locate all the documents you can find that address specific organizational performance metrics. What measurable promises have been made at the administration level (such as, government-wide goals for performance-based acquisition)? How about at the agency level? Has the administrator of your agency committed to reducing processing time to better serve citizens? Does that affect your project? What is the program trying to achieve that your project affects? What performance objectives exist for your project?

3. Individual Performance Evaluation Framework

Now, see what you can find out about how the goals and objectives set for the organization are translated into individual (personal) performance evaluation and measurement plans. What are your performance objectives? How about those of your boss? Those of your boss's boss? What relation (if any) do they bear to what you know about organizational goals, objectives, and metrics?

4. Contract Performance Evaluation Framework

Finally, take a look at the contract if one has been awarded—or the solicitation (final or draft) if a contract has not been awarded. Do (or will) the contract's objectives align with what you learned? Are there ways to improve alignment within the scope of the contract (or by redrafting or modifying the solicitation)? Can the way in which you manage the contract improve alignment? Are the metrics well chosen to measure progress? If not, can they be improved? Can effective contract management improve performance measurement and management over time? If a contract has not yet been awarded, consider how the contract can be aligned with the other frameworks you have studied.

What tools and techniques can be used to successfully implement strategic linkage in an organization?

The input you gathered on goals, objectives, measures, and metrics is likely a collection of somewhat related parts, rather than a tightly integrated whole. What tools and techniques can you use to create unity from the pieces?

1. Map the Organization's Goals and Objectives

From all the materials you gathered, create first a hierarchical map of the organization's cascading goals and objectives and their interrelationships. The following example, published by the General Services Administration's (GSA's) Office of Governmentwide Policy in 1998, illustrates the statutory foundation at that time for the Department of Education's Office of Student Financial Aid (now Office of Federal Student Aid). The statutory foundation[4] should be expanded with program, project, and contract goals and objectives to create the overall foundation for strategic linkage.

Table 5-1		
GPRA—Agency Performance Goal, Objective, and Strategy	ITMRA—Program Performance Measures	FASA—Acquisition Cost, Performance, and Schedule Goals
Goal: Ensure access to post-secondary education and lifelong learning.		
Objective: Post-secondary student aid delivery and program management are efficient, financially sound, and customer responsive.		
Customer satisfaction ratings among students, parents, and post-secondary institutions participating in the student aid programs will increase to 90% by 2001.		
Evaluation of contracts for major OPE financial aid systems will indicate that the government and the taxpayer are receiving "better than fully successful" performance (including quality, cost control, and timeliness).		
By September 1998, ED will have a complete system architecture developed for the delivery of federal student financial aid; implementing this design will improve customer service and increase control over federal costs.		
	Reduce by at least a third the amount of hard copies of materials that now must be printed and mailed.	
	Enable applicants for Federal aid filing electronically to have their eligibility determined in four days, cutting in half the current processing time.	
Core Strategy: An integrated, accurate, and efficient student aid delivery system, including (in part) the supporting strategy to increase the community's use of the Department of Education's web site as a principal source of financial aid information, programmatic and technical publications, and software.		
		Cost Goal: $___,___.

Source: "A Guide to Planning, Acquiring, and Managing Information Technology Systems," GSA Office of Governmentwide Policy, 1998.

2. Map the Organizational Performance Framework

The next step is to map what you learned about your organization's performance framework. Acquisition Solutions is a performance-based organization and has established an organizational performance framework to ensure that "believing is being." The company uses a balanced scorecard that establishes goals and objectives and metrics at the corporate level that relate to client satisfaction, employee satisfaction, process and quality, and financial performance and growth.

3. Understand the Individual Performance Evaluation Framework

At Acquisition Solutions, the goals and objectives and metrics set at the corporate level cascade throughout the organization to offices and individuals, creating strategic linkage and alignment among employees about what the company is seeking to achieve. This works in government, too. When one of Acquisition Solutions' senior leaders worked for the General Services Administration in the 1980s, the personal performance evaluation framework cascaded down three levels, from the Senior Executive Service-level service head to her as first-line supervisor and then to all her staff. All had insight into how (at least) four levels of management would be evaluated, and this created alignment of interests and intent.

4. Assess the Contract Evaluation Framework

If you apply the discipline of strategic linkage early in the acquisition process, you will have the opportunity to shape the contract evaluation framework—from evaluation and selection criteria, to in-process reviews, to incentive and award fees (and the like), to past performance evaluations—all in tight conformity with organizational performance objectives.

If you apply strategic linkage after the solicitation has been released or (worse) after contract award, your options are more limited ... but there are always options. For example, one client agency had a contract in the late 1990s that was heading toward failure and a termination decision. This was in no small part because there was no strategic linkage, nor was there alignment between the government's and contractor's interests. But neither the government nor the contractor wanted the contract to fail. Acquisition Solutions worked with parties from both "sides" to negotiate a contract transformation that created linkage and alignment and turned performance around.

5. Overlay the Maps, Identify the Gaps, and Create the Master Linkage Strategy

Once you understand the organization's goals and objectives, the approaches taken to organizational and personal performance evaluation, and the contract performance evaluation framework, you can overlay the maps and identify the gaps. What is revealed can be quite interesting.

For example, Acquisition Solutions once supported the transformation of an award fee task order into a revised structure that allowed for better alignment to and management of performance, namely, an incentives and disincentives framework in a performance-based environment. A tremendous amount of energy had gone into establishing a complex measurement process; however, it failed to focus on strategic alignment.

The recommendation for transformation arose from the following pertinent facts:

- The task order had more than 100 performance measures and metrics. It could be more effectively measured and efficiently managed with fewer but more essential performance measures and metrics.
- Many of the measures and metrics were, in fact, mandatory requirements that were inappropriately tied to an award fee plan. Why give a fee award for a requirement?
- The core (objectives-based) performance measures and metrics could be easily measured, but were tied into a subjective award fee evaluation instead of a more appropriate incentives and disincentives strategy.

The acquisition team and task managers reviewed and evaluated every performance measure and metric to determine its initial intent, validity, and relative importance to the mission. As a result, the performance measures and metrics were reduced from more than 100 to approximately 16 with the highest correlation to mission objectives and criticality and to program success.

The point is ... it's never too late to achieve strategic linkage and alignment. You need to identify the gaps, develop a strategy for creating or improving strategic linkage, and then execute and master the implementation of that strategy.

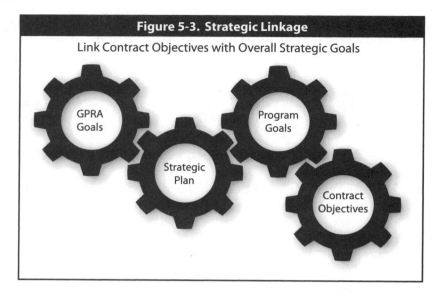

Figure 5-3. Strategic Linkage

Link Contract Objectives with Overall Strategic Goals

GPRA Goals

Program Goals

Strategic Plan

Contract Objectives

Case Study: Strategic Linkage of the FEMA Map Modernization Project

FEMA Mission: Lead America to prepare for, prevent, respond to, and recover from disasters.

FEMA Strategic Goals:
- Reduce loss of life and property.
- Minimize suffering and disruption caused by disasters.
- Prepare the Nation to address the consequences of terrorism.
- Serve as the Nation's portal for emergency management information and expertise.
- Create a motivating and challenging work environment for employees.
- Make FEMA a world-class enterprise.

In its business case, FEMA concisely but completely, in just over one page, described how the Map Modernization project directly links to each one of these goals. For example, demonstrating linkage to the first goal, the business case indicated, "Accurate, up-to-date flood hazard maps are key to achieving this goal. . . . Many communities are using outdated maps on which to base critical decisions for protection of life and property." Linkage to the second goal was addressed, in part, by indicating, "With accurate hazard data, state and local agencies can take precautionary actions to minimize multi-hazard impacts, develop flood hazard warning systems, and support other state agency efforts."

6. Maintain a Culture of Performance with Empowerment and Accountability

Strategic linkage depends on a culture of performance, empowerment, and accountability. A successful system of accountability results in the managers and employees from multiple sectors "buying in" and assuming responsibility for some part of the process. Empowerment and accountability are strategies are used by world-class organizations to cement strategic linkage for performance success.

Employees are most likely to meet or exceed performance goals when they are empowered to make decisions and solve problems in areas for which they are accountable. In many ways, accountability is analogous to a contract between manager and employee, with the manager providing a supportive environment and the employee providing results. The performance goals of an organization represent a shared responsibility among all its employees, each of whom has a stake in the organization's success. A critical challenge for private and public organizations alike is ensuring that this shared responsibility does not become an unfulfilled responsibility.

Accountability helps organizations meet this challenge. According to one participant in the National Performance Review's June 1997 benchmarking study, "Serving the American Public: Best Practices in Performance Measurement,"[5] "the system we use is a closed loop . . . responsibility is attached to authority resulting in accountability." Another commented that "you can only hold employees accountable if they have control."

7. Develop at All Levels Well-defined Roles and Responsibilities

Responsibility and accountability for results must be clearly assigned and well understood. While senior leadership ultimately is responsible for establishing the strategies, goals, and performance objectives for the coming year, clarity around roles and responsibilities—as well as performance expectations—is essential.

Thus, once objectives and goals are established, they should be passed down to the next organizational structure. This provides direction from and hierarchical linkage to the plan's highest level. The hierarchical linkage attribute occurs because of the pass-down process for the plans at each succeeding level.

At each level, the plan from the next higher level is used to develop a plan for that level. Each group simply uses the strategic objectives from above as the foundation for its objectives and goal—with the cross link being the roles and responsibilities as well as the group's expertise or technical strengths. In this way, the organization's critical issues and plans filter down through the organization, with *each level and unit and individual* contributing where it appropriately and most effectively adds value. At each succeeding level, strategies are owned, expanded, and turned into implementation plans that contribute to reaching the objective and the overall goal. This cascading effect is a very important step in empowering and uniting the organization.

8. Create a Quality Assurance Surveillance Plan

An important contract document for strategic linkage is the quality assurance surveillance plan (QASP). It defines success for each objective (along with sample measures) and establishes a common understanding among the government and contractor team members.

The QASP recognizes the contractor's responsibility for quality control, recognizes the government's responsibility to ensure it, and includes measurable inspection and acceptance criteria corresponding to the performance standards specified in the contract (generally in the quality assurance plan, preferably submitted by the contractor). The focus of the QASP must be on level of performance and *not* on the manner in which the contractor performs. The following provides a partial example of a QASP for help desk activities.

Table 5-2					
Desired Outcomes *What do we want to accomplish as the result of this contract?*	**Required Service** *What task must be accomplished to give us the desired result?*	**Performance Standard** *What should the standards for completeness, reliability, accuracy, timeliness, quality, and/ or cost be?*	**Acceptable Quality Level** *How much error will we accept?*	**Monitoring Method** *How will we determine that success has been achieved?*	**Incentives/ Disincentives for Meeting or Not Meeting the Performance Standards** *What carrot or stick will best reward good performance or punish poor performance?*
1) Customers calling the help desk shall be able to contact a support staff member from 8:00 a.m. to 5:00 p.m., M-F.	The help desk shall be adequately staffed, with a sufficient number of incoming lines to handle potential trouble calls.	99% of calls are answered on the customer's first attempt.	99% of calls are answered on the customer's first attempt.	Survey customers and evaluate feedback. Inspect call logs. (Trend analysis.)	+/- .5% of total monthly price.
2) Calls are answered promptly by help desk personnel.	The help desk shall be adequately staffed, with a sufficient number of incoming lines to handle potential trouble calls.	Calls are answered within 20 seconds or a voice mail can be left; calls shall be returned within one hour of receipt.	Calls are answered within 20 seconds or a voice mail can be left; calls shall be returned within 30 mins. for L1 customers and 60 mins. for L2 customers.	Random sampling of call activity logs, showing time of receipt of call and call return time. Random surveillance of actual operations. (Trend analysis.)	+/- .5% of total monthly price
3) Time to resolve customer problem or answer question is as short as possible; the need to dispatch personnel is minimized.	Time to resolve problems or answer questions is within the time frames set forth in the SOW or in the Service Level Agreement (SLA).	96% of calls received are resolved within 1 business day.	96% of calls received are resolved within 1 business day.	Random sampling of call activity logs, showing time of receipt of call and closeout of trouble tickets. (Trend analysis.)	+/- 1% of total monthly price

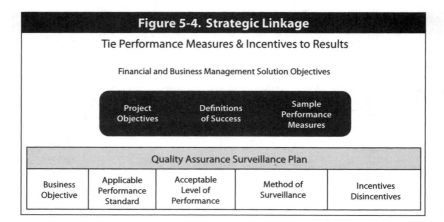

Figure 5-4. Strategic Linkage

Tie Performance Measures & Incentives to Results

Financial and Business Management Solution Objectives

| Project Objectives | Definitions of Success | Sample Performance Measures |

Quality Assurance Surveillance Plan				
Business Objective	Applicable Performance Standard	Acceptable Level of Performance	Method of Surveillance	Incentives Disincentives

9. Align Performance Measurement for Strategic Linkage

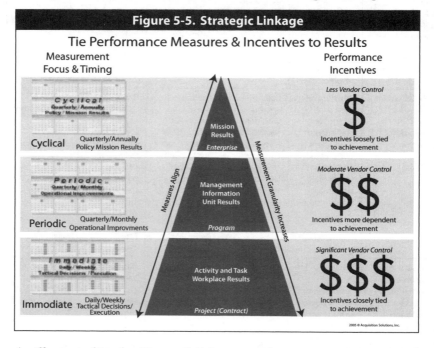

Figure 5-5. Strategic Linkage

Tie Performance Measures & Incentives to Results

Measurement Focus & Timing

Performance Incentives

Cyclical — Quarterly/Annually Policy Mission Results

Periodic — Quarterly/Monthly Operational Improvments

Immodiate — Daily/Weekly Tactical Decisions/ Execution

Mission Results — Enterprise

Management Information Unit Results — Program

Activity and Task Workplace Results — Project (Contract)

Measures Align

Measurement Granularity Increases

Less Vendor Control
$
Incentives loosely tied to achievement

Moderate Vendor Control
$$
Incentives more dependent to achievement

Significant Vendor Control
$$$
Incentives closely tied to achievement

2005 © Acquisition Solutions, Inc.

As illustrated in the Figure 5-5, because the contract, project, and program objectives are tied to the mission, goals, and objectives of the organization, performance measures should align both from bottom (contract/project) to top (enterprise/organization) and top to bottom, with increasing granularity of measures, information, and reporting.

Contractor performance incentives should be tied proportionately to the right strata. Contractor measures, metrics, and incentive structures should take into consideration the degree of control the contractor has over the outcome. The more control the contractor has over the results of the task, the more closely tied incentives should be. For example, for the FEMA Map Modernization project, the higher the level of the measure (e.g., "reduce the loss of life and property"), the less control the contractor has over that outcome. As a result, a smaller proportion of the contractor's incentive should be based on that outcome. The more granular the measure (e.g., "produce accurate flood hazard maps"), the easier it is to tie it to the contractor's performance.

Remember, too, that measures must be "SMART," an acronym that frequently appears when commentators discuss performance measures. It means that performance measures need to be Specific, Measurable, Agreed to, Realistic, and Timely. In general, a good measure is accepted by and meaningful to the customer; tells how well goals and objectives are being met; is simple, understandable, logical, and repeatable; shows a trend; is unambiguously defined; allows for economical data collection; is timely; and is sensitive.[6] Above all, however, a good measure drives appropriate action.

It is important to provide some incentive tied to mission results. This fosters buy-in and a "we-are-all-one-team" outlook, which is a best practice of managing for results. Most everyone has heard about incentives and disincentives, but the power comes from performance measurement systems that are positive, not punitive. The most successful performance measurement systems are not "gotcha" systems, but learning systems that help the contractor, team, or organization identify what works and what does not, so as to continue to improve.[7]

What are the key elements to consider in performance measurement? The Government Accountability Office (GAO) concluded that there is a direct linkage between individual performance and organizational success. In a report[8] published in March 2003, GAO identified nine policy actions that successful organizations pursue:

- Align individual performance expectations with organizational goals
- Connect performance expectations to crosscutting goals

- Provide and routinely use performance information to track organizational priorities
- Require follow-up actions to address organizational priorities
- Use competencies to provide a fuller assessment of performance
- Link pay to individual and organizational performance
- Make meaningful distinctions in performance
- Involve employees and stakeholders to gain ownership of performance management systems
- Maintain continuity during transitions.

To the extent possible on individual contracts, best-in-class organizations provide performance incentives or other recognition not only for the contractor but for the government team as well. This helps forge a "we" attitude and mind-set in contract performance, rather than "us versus them," and builds in an incentive to collaborate with the contractor.

10. Implement Recognition and Reward Programs

Recognition and reward programs, commonly used in the private sector to recruit and retain employees, also can be implemented in the government. The Office of Personnel Management cites regulations at 5 Code of Federal Regulations (CFR) 451 that "allow agencies the flexibility to recognize employees for a wide variety of accomplishments, as well as to create incentives, using four forms of awards: cash, time off, honorary awards, and informal recognition." [9]

What is the difference between recognition and incentives? Recognition provides after-the-fact reinforcement for specific types of performance or accomplishments and signals what the organization values. Incentives focus employee efforts on the organization's goals and often promise specific rewards to those employees who help significantly to achieve them. Federal award programs have tended to reward performance and accomplishments (that is, provide recognition) rather than encourage new or improved performance (that is, serve as an incentive). Performance-based project managers should analyze their objectives, measures, work structures, and cultures to determine the types of recognition and rewards the might work best. The award regulations provide enough flexibility to allow a wide variety of approaches and methods for recognizing employees and providing incentives. [10]

11. Implement Performance Measurement Best Practices

As defined by the National Performance Review's benchmarking study"[11] and enhanced by Acquisition Solutions' experience, the following are some best practices:

- *Ensure a narrow, strategic focus.* The measures and goals an organization sets should be narrowly focused to a critical few. It is neither possible nor desirable to measure everything. In addition, mature performance measurement systems are linked to strategic and operational planning.

- *Implement customer-driven strategic planning.* Measurement should concentrate on items that can be traced through business unit performance plans to the entity's strategic vision. If a measure and its corresponding data requirements cannot be linked back to strategic planning, they should be considered for immediate de-emphasis or elimination. This frees organizations from "rescue initiatives" in areas that produce little value and, equally important, avoids data overload.

- *Measure the right thing.* Before deciding on specific measures, an organization should identify and thoroughly understand the processes to be measured. Then, each key process should be mapped, taken apart, and analyzed to ensure: (1) a thorough, rather than assumed, understanding of the process, and (2) that a measure central to the success of the process is chosen. In some cases, targets, minimums, or maximums are defined for each measure.

- *Be a means, not an end.* In a best-in-class organization, employees and managers understand and work toward the desired outcomes that are at the core of their organization's vision. They focus on achieving organizational goals and use performance measures to gauge that achievement, but they do not focus on the measures per se. Performance measurement is thus seen as a means, not an end. Focus on the goal, measure the results, and don't focus on the measurement.

- *Set up the data collection process to match the needs of your organization.* An organization must develop performance measures that complement its culture, size, mission, vision, organizational level, and management structure, as well as its goals and objectives. Regardless of size, sector, or specialization, organizations tend to be interested in the same general aspects of performance:
  - Customer satisfaction

- Employee satisfaction
- Stakeholder satisfaction
- Internal business operations
- Financial considerations

- *Measure both efficiency and effectiveness of the organization.* It is essential that we measure both the efficiency and effectiveness of an entity. By efficiency, we mean, *Are we doing things right?* Are the results and objectives in meeting the project goals being accomplished in a cost-effective manner?· Important efficiency questions are:
  - Was the effort completed on time and within budget?
  - How much was produced?
  - How many resources were required?

- By "effectiveness" we mean, *Are we doing the right things—and achieving results that matter?* Important effectiveness questions are:
  - Has the organization achieved its mission and goals? If not, what part of the mission requires bolstering?
  - Are end users of its products and services satisfied customers? If not, what areas require improvement?
  - Was the work of high quality? If not, where must it improve to satisfy customers?

What are the desired outputs of the strategic linkage process?

As defined by the National Performance Review's benchmarking study, the desired outputs center on turning data into information, disseminating information, and acting on it.[20]

1. Turning Data into Information

Data analysis in performance measurement is the process of converting raw data into performance information and knowledge. The data that have been collected are processed and synthesized so that organizations can make informed assumptions and generalizations about what happened. They then can compare what actually happened to what they thought would happen, decide why there might be a variance, and determine what corrective action might be required.

Everyone needs information, but not everyone knows what to do with raw data. So, frequently in world-class organizations, in-house quality staff or outside contractors analyze the data used to measure performance. Some organizations provide data directly to managers, or to the relevant business unit, for analysis. At one world-class organization, data analysis takes the form of "cross talks" between organizational units jointly reviewing performance results. To ensure everyone can use and understand data and its analysis, some organizations train their workforces in basic analytical methods. Such an evaluative culture is promoted by engaged executive leadership and often nurtured by a cadre of analysts helping business units understand and interpret their data.

User information needs differ. Different levels of an organization will use different pieces of the analyzed data. The critical users of performance information are decision makers, both on the front lines and in the executive suite, not the analysts who convert data to information. In fact, the goals of analysts are not always synchronized with the goals of the decision makers. As a rule, decision makers need information that is timely, relevant, and concise; analysts tend to value products that are thorough, objective, and professionally acceptable. The successful resolution of this internal tension is a sign of a world-class performance information system.[13]

2. Disseminating Information

Communication is crucial for establishing and maintaining a performance measurement system. It should be multidirectional, running top-down, bottom-up, and horizontally within and across the organization.

Effective internal and external communications are the keys to successful performance measurement and to maintaining strategic linkage. Communication should extend to employees, process owners, customers, and stakeholders. It is the customers and stakeholders of an organization, whether public or private, who ultimately will judge how well the organization has achieved its goals and objectives. And it is those within the organization entrusted with and expected to achieve performance goals and targets who must clearly understand how success is defined and what their role is in achieving that success. Both organization outsiders and insiders need to be part of the development and deployment of performance measurement systems.

3. Acting on Information

Dr. Bob Frost, who directs the consulting firm Measurement International, acknowledges the challenges of the alignment process. "For a measurement system to become 'the way we do business around here,'" Frost says, "three things have to happen:

■ People have to know about it,
■ People have to care about it,
■ People have to be able to act on it."

In a best-in-class organization, employees and managers understand and work toward the desired outcomes that are at the core of their organization's vision. They focus on achieving organizational goals, using performance measures to gauge goal achievement, but do not focus on the measures per se. Performance measurement is thus seen as a means, not an end. And done correctly, the processes forces the evolution to a learning organization that continually adjusts and improves performance.

SUMMARY

Organizations (both public and private) need to tap into their data to discover what is working and what is not. They will thrive if they adopt an attitude of inquiry, not embattlement, towards linking goals and performance measurement.

When organizations make measurement information readily available at all levels of the organization, as well as to stakeholders and customers, when tools are provided to support widespread analysis by more people across the organization, and when an attitude of interactive inquiry prevails, great things begin to happen:

■ Previously unrecognized problems are found
■ Known problems are better characterized and understood
■ More precise, effective, and cost-effective treatments are noted and implemented

Finally, well-designed measurement systems, strategically linked, provide relevant information to decision makers when they need it. They inform policy decisions and consumer choice, and ultimately improve contract, project, program, and organizational performance.

FIVE

In the next chapter, we discuss the Third Discipline of Performance-Based Project Management—the practice of governance principals and tools.

QUESTIONS TO CONSIDER

1. How important is strategic linkage to performance results?

2. Is accountability an overall strength of your organization and team members?

3. What tools and/or techniques does your organization use to define contract, project, and program success?

4. Do you measure both effectiveness and efficiency of a contract, project, and program?

5. How effectively does your organization recognize and reward outstanding performance?

6. Does your organization use a performance balanced scorecard to plan and measure success?

Endnotes

[1] Acquisition Directions™ *Advisory*, "Performance-Based Acquisition Requires the Six Disciplines of Performance-Based Management," Acquisition Solutions, May 2004.

[2] William Fonvielle and Lawrence P. Carr, "Gaining Strategic Alignment: Making Scorecards Work," *Institute of Management Accounting Quarterly*, Fall 2001, *http://www.imanet.org/publications_maq_back_issues_fall2001.asp*.

[3] OMB Circular A-11, "Exhibit 300, Capital Asset Plan and Business Case," is the budget justification and business case for a major investment.

[4] The Government Performance and Results Act, the Information Technology Management Reform Act, and the Federal Acquisition Streamlining Act.

[5] National Performance Review, Benchmarking Study, "Serving the American Public: Best Practices in Performance Measurement," June 1997, *http://govinfo.library.unt.edu/npr/library/papers/benchmrk/nprbook.html*.

[6] Ibid.

[7] Ibid, n. 5.

[8] "Results-Oriented Cultures: Creating a Clear Linkage between Individual Performance and Organizational Success," General Accounting Office Report No. 03-488, March 2003.

9   Office of Personnel Management website: https://www.opm.gov/perform/articles/207.asp.

10   Ibid.

11   Ibid, n. 5.

12   Ibid, n. 5.

13   Ibid, n. 5.

CHAPTER 6

THIRD DISCIPLINE
— GOVERNANCE

By: Anne Reed

INTRODUCTION

The term "governance" takes on different meanings depending on the context. Citizens look to their governments to govern by establishing and enforcing laws. In the corporate world governance tends to be discussed in terms of the roles and responsibilities of the boards of directors – to assure that shareholder interests are well represented. In the world of Performance-Based Project Management (PBPM), project managers usually refer to governance as a path to decision making.

> At its essence, governance is a framework that underpins the consistent development and application of policies and processes so that the people involved can have confidence that the outcomes are aligned with the stated goals.

For our purposes, we focus on governance from the perspective of a project that is commissioned by one or more buying organizations, entrusted to a performance-based project manager (you) and executed in whole or in part by a contractor.

Over the last decade, both industry and government have moved towards outsourcing services. This has created a need to readdress project governance so that the goals are clearly understood by all parties and the processes and policies are aligned to achieve the goal.

While never simple, now there are even greater complexities requiring the need for increased focus on governance of the project including:
- the governance of the acquiring organization
- the governance of the project or program itself – and the governance established by the contract(s) that support it, and
- the governance of the service provider(s).

Too often, there has been little recognition that the project managers must be sensitive to the needs of their supply chain, as well as their own management.

In both the public and private business sectors, the ground is littered with failed projects, especially Information Technology (IT) projects. While there are many contributing reasons, one of those that is often overlooked is the lack of clarity in the project governance model.

Personal Story

I remember sitting in on an evaluation of a program that was "in trouble." The project – a government project to modernize processes – was large and complex involving many interlocking sub-projects. There was a government manager and a systems integrator manager overseeing many sub-contractors. They were behind schedule, over cost and under intense scrutiny by senior management and oversight organizations. To try to better understand the causes underlying the program difficulties, a matrix of roles and responsibilities was created, focusing on key decision points. Project and sub-project leaders from both the government and the contractor organizations were interviewed individually. On each decision point, they were asked if they thought they had the authority to make the decision and if not, who did.

The answers were quite revealing. There was no common understanding of responsibilities and authorities. Some people assumed they had authority that no one else credited them with. Others expected decisions from people who did not themselves think they had the responsibility. There were key elements where <u>no one</u> believed they had approval authority and others where two or more people claimed the final authority.

Nearly everyone understands how a lack of clarity around decision making authorities creates frustration and delays. Project delays usually cost money – and they mean that the project goals aren't being achieved. In this chapter, we will focus on the importance and value of creating a project governance structure and the key principles and proven best practices to achieve project success.

WHAT IS THE GOVERNANCE PROCESS?

In order to effectively and efficiently implement a project governance structure, it is helpful to follow a proven process approach. Figure 6-1 illustrates the key inputs, tools, and techniques, and the desired outputs to create a successful project governance structure, which facilitates getting high performance results.

Performance-Based Project Management

Figure 6-1. Third Discipline – Governance Process

Key Inputs	Tools & Techniques	Desired Outputs
Principles of Good Project Governance • Reconfirm the Goal • Establish Common Values • Agree on Objectives & Perfomance Metrics • Recongnize the Constraints • Clarify Key Roles and Responsibilities • Know the External Influences	Governance Planning Governing Documents • Project Charter • Project By-Laws • Contract • Organization Chart • Responsibility Assignment Matrix	Governance Processes Establish Accountability Mechanisms Consistently Observe Processes Make Good Project Decisions in a Timely Manner Excalate and Resolve Project Issues Effectively Build a Strong Contractor/Program Relationship Achieve Project Goals

WHAT ARE THE PRINCIPLES (KEY INPUTS) OF GOOD PROJECT GOVERNANCE?

It is a simple thing to say you need a strong governance framework. But it is not always so simple to create one. The first step is simply to recognize that you need one. What next? Think in terms of key inputs, which together form the overarching principles of good project governance. The key inputs follow:

1. Reconfirm the goal

If the ultimate purpose of a governance model is to assure that the outcomes are aligned with goals, then surely it is important to validate that all involved share the same understanding of what the goal is. Restate the mission or objective for the project. What is the purpose to be achieved? Hopefully, the project plan and the contract will be clear about the objective. But sometimes there is no project plan or charter. And often contracts are fairly generic – or are all about "how" to accomplish the objective without being deeply grounded in the overall mission outcome.

Example: Your company's goal is to sell more cars quickly in order to make room for the new inventory expected in six weeks. To help achieve this objective, you need to contract with an advertising agency. There are three primary ways that you can frame the requirement:

Generic "boilerplate": Design and place car advertisements.

Specification-based: Take photographs of mothers and children in Company X mini-vans and use them in advertisements to be placed in the local newspaper.

Objectives-based: Conduct campaign to reduce current car inventory of xx to xx in six weeks.

While the first two achieve the objective of advertising cars, neither alone will achieve the goal of selling them quickly to make room for new inventory. That is the strength of the objectives-based requirement.

Knowing what the goal really is will help the ad agency understand the urgency of the requirement and may enable them to be more creative in proposing solutions to the challenge. For example, they may suggest expanding beyond newspaper ads to include radio or TV spots where they can leverage discounts with local stations – or recommending that you feature young people because it's graduation season.

Bottom line – don't assume that all parties already know what the goal is. When you kick off the project, even if you think it is simple, take the time to revalidate the goal. If the project is a large and complex then the need is even more compelling.

2. Establish common values

It never hurts to be open about the type of relationship that you want to build with your contactor and your colleagues. When you hit those inevitable bumps in the road, having a relationship rooted in common values will be invaluable. Building an environment of trust enables you to work together to overcome those unexpected challenges.

Examples of common values include:
- **Provide for open disclosure and transparency** – everyone involved needs to know what the rules of the road are and have appropriate insight into the project status.
- **Act with integrity and ethical behavior** – strive for an environment where the agreed upon processes will be honored and decisions respected

■ **Respect the interests of stakeholders** – assure that there is a communications mechanism to engage multiple interests.

■ **Create trust** – provide a means by which individual parts of the project organization can trust that the other parts are stepping up to their obligations and responsibilities.

3. Agree on Objectives and Performance Metrics

How will you recognize success when you achieve it? It is important to agree upon the measures of success. The contract that you entered into should include performance metrics. If it does, revalidate them. If it does not, identify what the contract requires in terms of objectives or deliverables, and then work with the contractor to establish how you will evaluate and report on the contractor's performance. Then think about how to cascade the metrics down to the responsible parties on your respective teams.

Being accountable for a function means that everyone involved needs to mutually agree to the expectations. And, if you are the project lead, you want to make sure everyone is pulling in the same direction. Most projects have a lot of interdependencies, so keeping the work streams aligned becomes critical to staying on course.

There are a number of ways to document expectations: contractual language, memorandums of agreement, charters, budget documentation, project plans, and individual performance plans. The sources and uses of these documents vary, but the key element is to make sure that incentives are aligned. The organization's goals, project objectives, team plans and personal incentives have to be in sync. Rewards need to be aligned with the project objectives.

4. Recognize the constraints

While every project leader will want to minimize them, the fact is there are legitimate constraints that impact project outcomes. It is best to recognize those parameters and build them into the process– or prepare risk mitigation strategies. What are some examples of how constraints might impact decision making?

■ **Availability of key decision makers:** You are at a key milestone and need a Board decision to proceed. Is it possible to poll the Board electronically, or will you have to wait until they are assembled for their regular bi-monthly meeting?

- **Safety:** You are building a military vehicle that needs to be able to move swiftly through hostile territory – but only heavy armor will protect the lives of the occupants. Who makes the call on the balance between speed and safety? Is there a safety standard that must be complied with?
- **Privacy:** Your challenge is to build an integrated data base on your personnel for use by managers and human resources professionals. How do you protect the data that should be kept private? And who decides how much privacy is enough?
- **Budget:** The budget for your project was set before the cost of gasoline went up. Now travel costs for the team are far exceeding expectations. Is the budget fixed by line item, or do you have the flexibility/authority to realign costs within your project budget?
- **Laws and regulations:** It probably goes without saying that you want to operate within the law, but the impacts may need to be acknowledged. For example, if you are designing a new car engine, your design team does not have the authority to waive federal emissions regulations.

On the other hand, sometimes people assume constraints where none exist. How many times have you heard: "We've always done it like that!" Try to create a process that allows people to question the constraints, but that also enables fairly quick resolution.

5. Clarify the Key Roles and Responsibilities

One of the main objectives of a governance model is to bring clarity to roles and responsibilities. Each project needs to have a designated leader and for large or complex projects there will be many roles to fill. Usually when a contractor bids on a project they provide information on the key personnel they expect to assign. At the same time, you will have people on your team to whom you turn for internal support.

Think through what the streams of work will be and how to most effectively create teams to support them. When there is a leader identified from your organization and also one from the contractor who is engaged to support, manage, or deliver the task, then make a particular effort to clarify who is accountable for the results anticipated and what the working relationships will be.

Make sure that every key function has a designated responsible party – either from your organization, your contractor team, or your multisector team. Sometimes these functions will be performed by people who are assigned full time to you and the project, but often they involve integrated project teams (IPT) drawn from disparate organizations and often performing their project work on a part-time basis.

Examples of key roles include: executive oversight, project and task leaders, financial management, communications, change management, knowledge management, technical support, subject matter expertise, quality assurance and administrative support. What might their responsibilities encompass?

Executive Oversight: Major projects should have executive sponsors and/or investment review boards that will be charged with making major decisions at key points along the way. If you don't already have an executive oversight model defined, consider asking a senior leader to serve as the executive sponsor.

The Executive Sponsor should be someone who has a stake in the outcome and is willing to:
- Be available to review the project status on a regular basis and give honest feedback
- Provide advice on managing external influencers
- Represent the business case for the project to other executives
- Assist in clearing corporate hurdles and approvals
- Provide support in negotiations for budget and staffing.

Boards and Committees

You should also consider what types of advisory boards and committees would increase the likelihood of project success. Essentially, you want to create boards and/or committees comprised of the people who can help make the project successful – or who can create stumbling blocks if they are not well informed/engaged. A variety of approaches can be taken with boards and committees, some of which are discussed below.

Advisory Review Board: Depending on the nature of the project, there may be some benefit to commissioning an advisory committee. The advisory group would serve to support the Project Manager and the Program Manager. The type of people you might include on this board would be people who represent some of the interest groups – or people who don't have a direct stake, but are impacted by the project. These are people that you want to keep informed and can help make sure that you at least understand their issues. Thus, this Board is advisory in nature in two critical ways: the Executive Sponsor informs and is informed. Depending on the project, this group might include people such as:

- The Executive Sponsor
- Financial/Budget Officer
- Chief Information/Technology Officer
- Contracting Officer
- Legal representative
- Other key stakeholders, such as end users, union representatives, community associations, and special interest groups.
- Executive leadership from the contractor

Just as there is a need for an executive sponsor in your organization, it is very important to have commitment from higher level management in the contractor's organization. This should be an individual who performs for the contractor a similar role within the company as the Executive Sponsor plays for the project. Creating a personal working relationship between executives in both organizations will provide a stronger probability for success.

Typically, this is not a decisional body. Its purpose is more to aid in communications and to build community/extended support. They would not convene frequently – perhaps at project kick-off and then at project mid-point, or semi-annually. However, your communication plan should assure that they are kept informed of key events on a regular basis. This is another opportunity to use a web-site dedicated to project performance.

Program Executive Board: Major projects may have a Program Executive Board in lieu of or in addition to the Advisory Review Board. This Board should include as many senior level stakeholders as practicable and should focus on high-level objectives and decisions. These are the decisions that have a significant impact on the direction of the program. They are decisions that go beyond the authority of the Project Manager (and even the Program Manager) to decide.

As a performance-based project manager, your responsibility is to assure that major issues are presented to the board and that the members have all the facts required to make decisions. If your project is one that cuts across organizations, this board becomes even more important.

For a large scale, mission-critical project, typical representation might include:

■ Executive management from each affected program organization;
■ Chief Financial Officer, Chief Information Officer, and Chief Acquisition Officer
■ Union representative
■ Attorney
■ Senior Executive(s) representing the Contractor(s).

While it is common to create boards from within the organization, it is strongly advised that the board also include at least one senior executive from your contractor's organization. I have seen some cases where the board was actually co-chaired by the program Executive Sponsor and the President or Senior Vice President of the company under contract to support and execute the project. This is a powerful mechanism for creating a strong and personal alignment among the key players.

Technical Review Committee: The performance-based project manager can benefit greatly by having a team of subject matter experts who form a Technical Review Committee. This team can be charged to review key deliverables or proposed technical solutions and provide advice. If there are differences of opinion on technical direction, it is good to have a forum where the issues can be aired openly and in a constructive framework.

Within defined parameters (usually defined in terms of impact to cost and schedule), this committee, chaired perhaps by your Technical Director, may be empowered to make technical decisions. For issues that cannot be decided at this level, the committee can be charged with presenting the nature of the problem, the alternative solutions considered and their pros and cons, and hopefully an agreed upon recommendation. Depending on the governance and decision-making structure, this presentation could be to the Executive Sponsor, performance-based project manager, Program

Executive Board, or other group assembled because of their roles and responsibilities.

Personal Story

When I was the Chief Information Officer at the U.S. Department of Agriculture, we needed to contract for a new Department-wide telecommunications system. I had already formed a departmental CIO Council comprised of the agency level IT Directors and CIOs to deal with department-wide information management issues.[1] Collectively, for this project, we established a technical review committee comprised of our telecommunications experts to advise us on what kind of system we needed and to evaluate proposals. The committee was chaired by my headquarters director of telecommunications.

After contract award, this same group of experts served to support the transition to a new service provider. By involving these experts early in a structured way, we got the benefit of their knowledge and we created a forum for healthy debate and discussion – since each agency had slightly different requirements. By working together, they were able to come to a recommendation that best suited the department as a whole – and were all invested in the decision so better ready to support the transition.

The fact is, I had the authority to sign a contract without any advice. And we probably could have gotten to contract award months faster. But I firmly believe that it would have been harmful had I tried to do so without using this governance/decision making model. Had we even approached the decision as one determined by me after receiving input from the department's CIO Council, we still would have missed some of the issues and been faced at the end with a lot of frustrated telecommunications specialists who would have felt marginalized and who could have stalled the implementation. And worst of all, the implementation would likely have failed since the contractor would not have been as aware of the range of technical challenges that were going to be presented.

Change Control Committee: Very few large scale projects go exactly as initially envisioned. When a circumstance arises that could have a significant impact on the schedule, the cost, or the defined solution, then there needs to be a way to bring those issues forward and get expeditious resolution. Programs can falter when issues are allowed to fester and go unresolved. And they can falter when major decisions are made by a single task leader without allowing appropriate input from project team members.

As a practical matter, most of the changes will involve a series of small adjustments – within the authority of the task leaders or the Project Manager. A few will be larger and may need to be reviewed first by the Change Control Committee before being raised up to the Executive Sponsor or Program Executive Board for final decision.

Usually the levels of authority are established in terms of specific cost, schedule or performance parameters. For example, the Change Control Committee will review all changes that—
■ increase the project's scope or planned functionality, or
■ increase the task cost by more than 10%, or
■ will cause the team to miss a previously defined critical milestone.

The Change Control Committee offers a way for the Project Manager to manage the mid-course corrections that are required. This committee should be chaired by the Program Manager. The contractor's project manager might serve as the "executive secretary," setting the agenda and assuring that the issues are appropriately staffed before coming to the committee for resolution.

Other members of the Change Control committee might include:
■ Senior task level project managers (from either the program staff or the contractor's designated managers)
■ Contracting officer
■ Program financial manager.

Even if the committee is composed of people who meet regularly for different purposes, it is important to formalize the responsibilities for this specific function. There should be a charter and all the participants need to know that this is the purpose for which they are being convened. Decisions need to be documented and, in order to maintain appropriate accountability, all those responsible for executing the decisions need to be formally advised of the change in their parameters.

Project Manager

Every project should have one person who is the acknowledged Project Manager. This person is accountable for managing the project so that, overall, it is meeting expectations for cost, schedule, and performance.

Simple? Maybe not. When the project is being accomplished by support from a contractor, then there may be two perceived project leaders, each of whom is accountable to their respective organizations. When this is the case, take particular care to clarify not just roles and responsibilities, but also expectations between the two individuals. Clearly delineate all of the threads of activity, noting who has direct responsibility for each. Then, establish a formal process of collaboration and coordination.

Task Leader

Many projects will have multiple streams of work underway, each of which will require a designated responsible person. Below, I identify a number of key functions that need to be addressed in performance-based project management. Depending on the scale of the project, each of these functions could have a separate leader, who may or may not be a full time designated individual.

Financial Management

Keeping track of costs is fundamental to any project. Know what your budget is and have a process for approving expenses and tracking them against budget. Understand what the requirements are for managing purchases, billing and other transactions.

In addition, recognize that the project is impacted by financial processes driven by your organization <u>and</u> by the contractor's organization. For example, if the contract requires the purchase of a large amount of equipment, the contractor's project manager will have to work through his or her own company's requirements to get the capital necessary to execute the transaction.

Communications

Communications shouldn't be left to chance. Think about how to organize and make available all of the information that is needed to support the project – even if it is only the phone numbers of all the task members. Consider how to keep management and other stakeholders informed of progress and what kind of team meetings should be held and how frequently. Often it is useful to set up a web-based workspace to provide a single place to capture project information and status reports. This site can also serve as a useful governance tool – by clearly documenting roles and responsibilities and approval processes. Communications is addressed in more detail in the next chapter.

Change Management

Many large projects have an impact on the organization's business processes and will ultimately require people to interact or perform their jobs differently. Sometimes the change required is of such a scale that cultural transformation efforts (Chapter 4) become advisable … or even necessary. Someone needs to be considering how best to implement the changes from a people perspective. Will people need to be reassigned to new roles? Will they need training? Consider the human impacts related to your project. Plan for this and know how to get relevant decisions made and communicated in a timely manner.

Knowledge Management

Increasingly, emphasis is placed on working within a knowledge-enabled environment. As your teams develop, create opportunities for learning before, during and after – and capturing the lessons so that they can be applied by others going forward. You may choose to introduce tools to support this function. It is often helpful to assign knowledge management coordination to one of the team members – who can be trained in facilitation and in use of the supporting knowledge tools and processes. Knowledge management is discussed in more detail in Chapter 3.

Technical Support

You've assembled the team members and found office space. But where are the telephones and computers? Are they to be provided by the contractor? What kind of support might be required from IT to provide the web-based "work area" for the team – or to provide access to information on your company server? Will new technology be required to support research needed for the project? You need to have someone focused on assuring that these functions are performed.

Subject Matter Expertise

Most projects require subject matter experts who understand the business function impacted by the project. These are the people who can make the call on whether or not the proposed project solution is workable within the parameters of the operating environment. If you are working on a Red Cross project to manage the blood supply, then you want to be sure you have someone on the team who knows a lot about how blood is collected, stored, and distributed. If you are developing software to support human resources processes, you need to have the wisdom and experience of an HR practitioner.

Quality Assurance

Quality should not be an afterthought; instead, quality processes (such as peer assists and peer reviews) should be incorporated into the project's processes and schedule. In performance-based acquisitions, the contractor will be working under a Quality Assurance Plan, while the government team will monitor performance using a Quality Assurance Surveillance Plan.

Administrative Support

It is also important to consider what kind of administrative support will be needed in support of the project and the tasks that will be supported. These can run the gamut from planning meeting and off-sites, supporting financial analysis, developing and monitoring the project using tools like Microsoft Project®, recording and transcribing minutes from meetings, and many other functions. Sometimes these tasks are performed by the government, sometimes by contractor personnel, and sometimes they are shared. By anticipating administrative support needs and establishing clear responsibilities for them, the project will run much more smoothly.

6. Know the external influencers

Don't forget the "external influencers." Part of your challenge is to determine how and when you need to engage them in the project. These are people who aren't on the project team but who may make decisions or take actions that will have an impact on the project. And remember that they will exist both for your organization and your contractor. What do I mean by this?

Say you are an IT manager for a University and you have been tasked with modernizing the core infrastructure. You decide you want to move to a managed services model – where the University no longer needs to buy all the IT equipment, but instead relies upon a service provider to refresh equipment as needed to keep the whole system functioning within agreed upon service metrics. This is a large project and could take more than a year to fully implement. You've clearly identified the project management staff, both from your staff and the contractor you have hired. But who else can have an impact on your ability to successfully execute the project on time and within budget?

On the University side, your system touches students, faculty, and staff – and maybe it affects life support systems in a hospital, a laboratory, or a greenhouse. Think about the "end-users" who will be affected by the change because they can and will give voice to concerns which may arise. If you are a public university, you may be dependent on the state legislature for on-going funding. Or perhaps there is a Board that is closely monitoring technology investments.

Personal Story

I recently saw one major project cancelled – not because it was over cost or behind schedule, but because an external interest group became concerned that the changes might negatively impact them. They successfully lobbied senior management and the project leaders never even knew what hit them. There is every chance that direct communication with the interest group representatives could have enabled the project leaders to understand and then mitigate the interest group concerns – or, at the very least, to have prepared the senior leadership to expect the challenge and address it. In this case, the consequences of not paying attention to the external influencers turned out to be catastrophic to the project.

Don't just consider the influencers in your own organization ... or the stakeholders that can affect your influencers. After contract award, your industry partner is likely also to have a lot of folks to coordinate with. If, for example, they signed on to build and provide a major IT system, when they put the bid in, they got corporate approval for the solution and the price bid. However, now that you are ready for (or in the midst of) project execution, they will need to negotiate with their fellow managers to free up the people needed and they will need approvals from their financial office – who in turn may need to coordinate with bankers to get financing. At the same time, they may need to coordinate with suppliers to assure that the equipment required for the project will be available in a timely manner and within the cost structure anticipated. While it is correct to assume that, because they are contractually bound, they are accountable, the fact is that behind the scenes there are a lot of approvals still required to execute; the time needed to execute those processes needs to be taken into account.

While not part of the project team, there will be many players who will have influence on the success of the endeavor. It pays to think through who they are and how best to manage them. Working together − Project Manager and contractor − to anticipate all of the contingencies reduces risk, makes for a better partnership, and increases the probability of program success.

WHAT ARE THE TOOLS AND TECHNIQUES USED IN GOOD PROJECT GOVERNANCE?

These can be categorized in three broad areas: the governance plan, governing documents, and governing processes.

1. Governance Planning

You've thought through all the project players and their roles and responsibilities. Now how do you create the rules of the game? A lot depends on the size and complexity of the project. There is no point in creating an elegant solution to a simple challenge. On the other hand, you will not be well served by undue simplicity when the task is complex. For discussion purposes, let us assume that you are operating within a large and multifaceted organization and your assignment is mission critical and anything but simple. What must your governance plan address?

Existing Framework

First, you need to consider the normal approval channels that exist within your program organization. Occasionally, you will have a project where you can by-pass the normal channels and establish a unique project structure, but the reality is that most projects are executed within an already existing framework − one that you will need to plug into. Typical approval frameworks include those related to budget planning, acquisition planning, and investment review boards. If you have been recently recruited to lead a new initiative, then find a mentor or an assistant who is respected by the organization and can help you learn how to best work the system.

Within the context of your organization's operating processes, you should look to establish a governing approach that will give you the greatest chance for success. Consider how to form a cascading set of governing bodies that provide appropriate representation to assure that the decisions made are enforceable.

The Players

In considering the principles (key inputs) of good governance, you have thought through key players and their roles and responsibilities extending to executive sponsorship, boards and committees, and functional areas, such as communications, financial management, knowledge management, and much more. The results, which form the governance framework, are part of governance planning and will be documented in the Project Charter.

Day-to-Day Operations

There are several models that can work for structuring the day-to-day project management in an environment where there is a Project Manager supported by contractors:

- Outsource the whole thing to the contractor, placing almost total accountability for results with the contractor
- Set the Program Manager up as Project Manager and use the contractors to provide key technical expertise and staffing or to be responsible for particular streams or work
- Create a partnership between the Project Manager and the contractor's project manager.

What is important is that everyone understands what the working model is and that roles and responsibilities are clearly defined

I've seen instances where the contractor expected to have full authority and accountability – only to find that the Project Manager expected to be running things day to day. And I've seen cases where the Program Manager (who also was the Project Manager) expected the contractor to take the ball and run independently, but the contractor couldn't get access to the data or people needed for interviews without more inside support. In both cases the project is put at risk because of a mismatch in expectations.

Whether staffed by the contractor or the program office, create an organizational structure that aligns with program outcomes and covers all of the necessary functions (discussed earlier). Make sure that each individual person understands his or her roles and responsibilities. Make sure that individual performance plans are established to reward performance.

2. Governing Documents

When it comes to governance — saying it's so, doesn't necessarily make it so — as evidenced by the example cited earlier where there was a complete lack of understanding by key parties as to what their individual decision authorities were. In order to clarify accountability and expedite implementation:

- document roles, responsibilities, and authorities;
- record decisions; and
- share the information with everyone affected.

There are a number of ways to accomplish this, some very formal and some less so. What are some the typical ways to document roles, responsibilities and authorities?

Project Charters

The first step is to establish a project charter, sometimes referred to as "Terms of Reference." This is a document that describes the purpose and structure of the project. Typically, a charter should:

- state the mission and vision,
- record the performance outcomes/metrics expected,
- establish expectations for scope and duration,
- provide for resources (people and dollars), and
- identify the key stakeholders and their roles and responsibilities (in broad terms).

It is this last piece that establishes the overall governance framework. For purposes of the Charter, the key stakeholders might include: the Executive Sponsor, the Program Executive Board (to include senior management representatives of responsible/affected organizations), and the Project Manager. The charter may also commission the Advisory Committee.

Project By-Laws

Aligning with the overall project charter, you should establish By-Laws for the Program Executive Board. These By-Laws will cover:

- Purpose
- Membership
- Roles and Responsibilities
- Schedule of Meetings

A template for By-Laws is provided in the Appendix.

Contract

The contract that you enter into with your services contractor should be aligned with the program charter in terms of mission and vision, performance expectations, scope and cost. With respect to roles and responsibilities, your contractor may be engaged to perform many of the functions and will therefore assume some of the key roles.

In the post-award environment, the contractor becomes a stake-holder. You should stop to reevaluate the roles and responsibilities identified in the Charter to determine if it is appropriate to recognize and include the contractor. For example, you may consider placing an executive level representative from the contractor's organization on the Executive Board.

Organization Chart

The organization chart is a quick way for people to understand the various functions and, in broad terms, the hierarchy of decision making. It identifies the key functions, the individuals responsible for those functions, and their relationships. However, the chart may or may not capture the nuances of cross functional teams – or the blended responsibilities of the contractor team and the program team.

Below is one example of an Organization Chart.

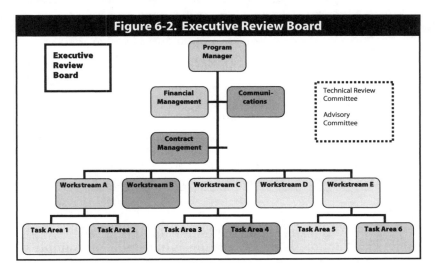

Figure 6-2. Executive Review Board

In this example, some areas represent work areas that are full time assigned to the Program Manager, some areas are all performed under contract and managed by the Project Manager (who is a contractor) and some areas reflect responsibilities that may be provided by people who are "matrixed" in (support provided from elsewhere in the organization). The chart is driven by function performed and the decision making hierarchy.

Responsibility Assignment Matrix

The Responsibility Assignment Matrix[2] is a simple way to assure that each of the project functions are clearly assigned to specific team members – and that each team member understands the expectations of them in terms of responsibility.

One common approach is based on the RASCI format. RASCI stands for Responsible, Accountable, Support, Consult, Inform. Some also add the responsibilities: Verifies and Signs. The table below defines each of these roles.

Table 6-1	
Role	**Definition**
Responsible	Conducts the work/owns the challenge. There should be only one R. If multiple R's are listed, then the work needs to be further subdivided to a lower level.
Accountable	Approves the completed work and is held fully accountable for it.
Supportive	Provides additional resources to conduct the work or plays a supportive role in implementation.
Consulted	Has the information and/or capability to assist in completing the work. Two-way communication (typically between R and C).
Informed	Needs to be informed of progress and results. One-way communication (typically from R to A).
Verifies	Checks the work to ensure that it meets all defined criteria and standards.
Signs	Signs off on the completed work.

The RASCI approach recognizes that for each function there are many parties who play a part in execution. This approach works equally well at both high and low levels of detail. An example of a RASCI matrix follows.

Figure 6-3. Responsibility Assignment Matrix

Task Description	Sponsor	Business Owner	Business Program Mgr	Process Manager
Identify missing or incomplete policies		R	A	R
Establish Policies as necessary and ensure adoption globally		A	R	R
Completion of necessary Policies		R	A	R
Document Policies as appropriate		R	R	A
Approve Policies	A	C	I	I
Communicate Policies as required		A	I	I
Ensure Policies are compatible with standards and best practice		R	R	A
Escalate non standard or missing policies	R	R	R	A
BP Sponsor with agreement from BPB colleagues decides on exception or not	A	I	I	I

WHAT ARE THE DESIRED OUTPUTS OF THE GOVERNANCE PROCESSES?

There are many processes that support governance, and they are employed in greater or lesser rigor, depending upon the size and scale of the project. Following are some representative governance processes.

1. Establish Accountability Mechanisms

An important part of governance is accountability – accountability for program performance, contractor performance, and personal performance. Build into the governance model mechanisms to assure accountability, for example:

■ Record decisions and document performance outcomes.

■ Take minutes of Board meetings other decisional forums (e.g., Technical Review Committee) and establish a formal means of approving and sharing the minutes.

■ If not formal minutes, capture the key "take aways" and action items that emerge from team meetings and share them via email to assure that everyone has the same understanding.

■ Establish forums for regular performance reviews (program and personal) based on program outcomes against plan.

■ Provide for status reports and assure that they are shared in whole or at a summary level with the entire team as well as with the Program Executive Board, Advisory Committee, and other governance groups.

Accountability is linked to performance and to transparency. People cannot respond to or act on what they do not know.

2. Consistently Observe Processes

Consistency in management approach is key to smooth implementation. Once you have established a governance framework around individuals and teams, it is important to honor the model.

■ If the framework calls for decisions to be made at the level of Task Manager, make sure that the Task Manager makes the call. If you are senior to the task manager, do not succumb to the temptation to intervene (at least not visibly!).

■ If the Program Executive Board is supposed to meet quarterly, make sure the meetings occur.

■ If you establish processes of monthly reviews and weekly reports, make sure that the reviews occur and the reports are created – and that they add value.

A key part of the value is in creating a level of confidence in the consistency of routine: it enforces discipline.

3. Make Good Project Decisions in a Timely Manner

For large projects, it is particularly important to have well-defined communications methods. Recognize that communications will serve many purposes:

- Status Reporting to Senior Management
- Updating stakeholders on progress
- Informing team members of decisions.

Do not assume that the leaders are sharing the key information on an informal basis. Different people have different styles of communicating. For purposes of governance, it is important to communicate key decisions in a formal way.

4. Escalate and Resolve Project Issues Effectively

Foster an environment that rewards honest debate and an open exchange of ideas. However, recognize that there needs to be a formal process for achieving resolution in a timely manner. The Responsibility Assignment Matrix should make clear where the final decision authority rests. However, the individuals who have the authority should assure that there is a reasonable opportunity to debate ideas before coming to conclusions – which are then documented and communicated to all who may be impacted.

5. Build a Strong Contractor/Program Relationship

Recognize that your contractors are stakeholders and have a vested interest in the performance-based project outcome. On at least a quarterly basis, convene a meeting of key contractor and project and program leaders to simply discuss what is working well with the relationship and what is not. Inevitably, the multisector workforces created by contractor/program teams create cultural challenges. It is best to create a time and place to have a facilitated discussion to air issues or concerns. Do not confuse this with a review of the contractor's performance in meeting the contract or program objectives. Rather this is a time to raise and clarify issues related to concerns about decision making authorities or the work environment.

6. Achieve Project Goals

Clearly, the most important output of governance—indeed of project management overall—is to achieve the project goals. This is the essence of performance-based project management.

SUMMARY

Governance is a framework that underpins the consistent development and application of policies and processes so that the people involved in managing a project can have confidence that the outcomes are aligned with the stated goals.

In acknowledging or creating a decision making hierarchy, it is important to recognize that there are diverse roles and responsibilities. Some are linked directly to the project (e.g., executive sponsor, project manager) and others acknowledge the roles of internal and external influencers (e.g., employees, financiers, citizens or suppliers). Attention to the governance framework becomes increasingly significant as the size and complexity of projects increase and when contractor support is part of the picture.

The governance model not only identifies the people involved and their functions, it also addresses the standard processes that are used to enable decision making – involving meetings of boards or governing bodies. Additionally, there are some documents that can help in bringing clarity to the governance framework. They include organization charts, charters, by-laws, and responsibility matrices. Other tools involve creating common web-based worksites where information can be shared.

Grounded in a clear statement of purpose, a set of established values and a common understanding of roles, responsibilities, and processes, the governance framework creates a measure of transparency that enables people to access the information they need in order to make good and timely decisions. By this manner, they may be accountable for achieving program results.

In the next chapter, we will discuss the 4th Discipline of Performance-Based Project Management – the practice of successful organizational project communications.

QUESTIONS TO CONSIDER

1. Does your organization create effective project governance structures?

2. What tools and techniques does your organization use to facilitate project governance?

3. Does your organization make critical project decisions in a timely and accurate manner?

4. Are stakeholders involved early and often?

5. Has careful thought been given to the identification and inclusion of the influencers in governance? Or have influencers emerged mid-project and affected the project's schedule or success?

Endnotes

[1]   The Department of Agriculture has 19 separate agencies, including the Farm Services Administration, the Forest Service, the Food & Nutrition Service and the Economic Research Service. You can see how their needs might be different since their sizes, geographical dispersion & missions are quite disparate.

[2]   Project Management Institute.

CHAPTER 7

FOURTH DISCIPLINE – COMMUNICATIONS

By Shirl Nelson

INTRODUCTION

Communications involves identifying the content, medium, and frequency of information flow, as well as the organizational elements the information is intended to support. With any initiative or project that requires a lot of collaboration to implement, good communication before, during, and after can mean the difference between failure and success. Project communications should be planned with consideration for why and what to communicate (the message), to whom (the audience), and how (the channels and frequency).

In the world of performance-based project management (PBPM), developing and implementing a project communications plan, particularly for large, complex projects, is standard practice for managers who want to get results. Rather than leave communications to chance—or worse, risk *mis*communication—we advise that a communication strategy be thought through and a written communications plan be prepared. Among other benefits, thoughtful project communication will help you manage stakeholder expectations, facilitate decision making, communicate accomplishments, and address problems quickly. Two-way vertical and horizontal communication facilitates the incorporation of lessons learned into new management strategies.

The focus of project communications should be on improving performance by passing information up, down, and sideways, and, as a team, resolving problems.

In this chapter, we examine the characteristics of good project communications and provide some tools and techniques to help organizations achieve high performance results.

WHAT ARE THE COMPONENTS OF A GOOD PROJECT COMMUNICATIONS PROCESS?

Figure 7-1 illustrates the key inputs, proven effective tools and techniques, and desired outputs of our recommended project communication process approach to achieve high performance results. The remainder of this chapter discusses each of the items listed in the figure in more detail.

Figure 7-1. Project Communications Process

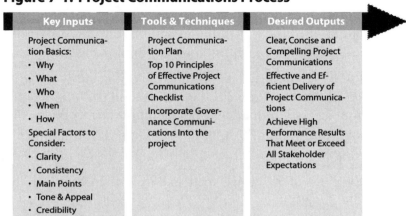

Key Inputs	Tools & Techniques	Desired Outputs
Project Communication Basics: • Why • What • Who • When • How Special Factors to Consider: • Clarity • Consistency • Main Points • Tone & Appeal • Credibility • Recipient Need	Project Communication Plan Top 10 Principles of Effective Project Communications Checklist Incorporate Governance Communications Into the project	Clear, Concise and Compelling Project Communications Effective and Efficient Delivery of Project Communications Achieve High Performance Results That Meet or Exceed All Stakeholder Expectations

WHAT ARE THE BASICS OF PROJECT COMMUNICATIONS?

There are five simple basics: why, what, who, when, and how. Let's examine each of these.

Why do I need to communicate? You might think that the primary purpose of project communication is to convey information. We do, of course, communicate to share information necessary to support a project, but that is only part of the picture. When managing projects or contracts for results, the ultimate goal of communication is to influence behavior. *We communicate to persuade people to take action toward our objectives.*

First, we want buy-in. We want our stakeholders, particularly those who might be on the fence or unsupportive, to get on board. And we want those who already are supporting us to help spread the message, so we need to articulate the message we want them to share. As project activity progresses, we want to prepare our stakeholders for inevitable changes and adjustments, to seek their input and minimize resistance.

We also communicate to bond the team. We want the team to share the objectives of the project so that they are supportive even when that means sacrificing parochial interests. Communication facilitates timely decision making. With information at hand, decision makers can more quickly reach consensus, and stakeholders can feel comfortable supporting their decisions.

Further, we need to communicate continuously to maintain alignment throughout the term of the project. Often, even project managers who pay attention to alignment at the launch of an initiative can lose focus. As time passes and problems and challenges begin to mount, stakeholders, particularly staff supporting a project, can lose sight of the purpose, goals, and objectives of a project. Turnover in project leadership or staffing calls for continual communication. Without it, time, performance challenges, and turnover can unravel the cohesiveness of the project team and their alignment with the goals of the project.

What is my message? Most project managers recognize the need to communicate the project's description and its goals and objectives to all stakeholders. The message should be tied to the objectives and compel the audience to think, feel, or act in a way that supports the project. Taking a cue from a professional marketing and communications group,[1] the message should:

- Show the importance, urgency, or magnitude of the issue
- Show the relevance of the issue
- Put a "face" on the issue
- Be tied to specific audience values, beliefs, or interests
- Reflect an understanding of what would motivate the audience to think, feel, or act
- Be culturally relevant and sensitive
- Be memorable.

The "what" also depends on the phase of a project. For example, throughout the project or contract life cycle, you should communicate next steps and any needed changes in direction. A best practice for horizontal communication during the course of a project is the use of the knowledge management techniques discussed in chapter 3. This includes seeking peer assists at the beginning of a project, conducting after action reviews and forming communities of practice during performance, and conducting thorough retrospective reviews at key milestones and at project completion. Lessons learned and success stories should be captured at every opportunity. This continual sharing of information enables timely decision making for course corrections and next steps, and highlights the critical information that should be shared with stakeholders up and down the chain and external to the team or organization.

While most project managers recognize the need to communicate a project's basic description, objectives, and plans, many overlook the benefits of communicating successes and achievements along the way. Yet, this is where the adage "Nothing succeeds like success" comes into play. Capturing the success stories and making them known is an excellent way to build support and to maintain or rejuvenate momentum for your project. Publicizing interim success also helps inoculate the project against criticisms that may arise later.

Case Study: FEMA Flood Insurance Program

In 2003, the Federal Emergency Management Agency (FEMA) awarded a contract with the objective of increasing by 5 percent a year the voluntary purchase of flood insurance policies. These would be policies for homes for which flood insurance is advisable but not mandatory, such as homes purchased prior to their locations being designated as flood zones. Flood insurance is expensive. Getting homeowners to voluntarily purchase policies is not easy. Good communications obviously played a key part in the performance of this contract, which required the contractor to communicate persuasively the benefits of flood insurance to homeowners in flood-risk areas. Our focus here, though, is the role good management communications played to keep the contract on target.

FEMA awarded this contract as a full performance-based project. The contractor had performance objectives, and the contract was managed through performance metrics. The first year of the contract, the target was not met, at least partly due to government actions rather than contractor performance. In the second year, FEMA kicked into full performance-based project management, and the number of policies purchased rose by more than 7 percent, and in the third year, the advertising campaign resulted in a more than 8 percent increase in the number of policies.

The project manager maintained support and enthusiasm for the project by communicating to stakeholders the reasons for the missed target in year one, then the achievements in years two and three. The contractor supported the success story by tracking which advertising channels worked and which didn't, making adjustments to its advertising approach, and communicating to the government the data and the basis for its performance decisions. Communicating the incremental achievements and successes kept the government from reverting to the traditional approach used in previous, unsuccessful contracts and resulted in FEMA's first successful flood insurance campaign.

You might wonder, Is it possible to communicate too much? Yes, we think it is. We often hear that more is better, but there is cost and risk associated with providing too much information, just as there is with providing too little. On the cost side, those requesting information can overburden others with demands for more data than can be used. Reports that are not necessary to support decision making are costly and counterproductive. On the risk side, those providing information can err by offering too much, burying the important message in an overabundance of detail. Sift through the excess and provide the nuggets. Achieving the right balance requires thinking through the purpose of and understanding the audience for the communication.

In a performance-based environment, reports should be limited to those that are meaningful and useful. The client and stakeholders should be asked what is useful to them and how they need to have that information presented. Avoid "wish lists" from different stakeholders requesting different information in different formats. Wish lists can lead to reporting demands that are unreasonable and unproductive. Whenever possible, contractors should be permitted to provide standard reports in a format producible by their existing systems. Examples of meaningful reports include:

- Regular reports on contractor performance
- Reports on internal program health
- Reports on program effectiveness in achieving agency goals and objectives
- Refined standard operating procedures
- Updates to program plans
- Reports on cost control, schedule accuracy, and quality of service delivery.

In all cases, reports should be concise, easy to understand, and tailored to various needs and audiences. When presented in a concise and easy-to-understand manner, information can be quickly digested and focus can be aimed at problem areas.

Who is my audience? Good communication reflects an understanding of the audience. This requires some analysis of who the communicator should be and who the audience is at any given time for a particular message. Over the course of a project or contract, there are likely to be many different audiences. A good communications plan, then, analyzes the multiple audiences and targets specific messages to each, based on the roles and interests of the stakeholders. Depending on the dollar value and complexity of the project, stakeholders might include not only internal audiences of managers, colleagues, and subordinates, but also such external audiences as oversight organizations, citizens' groups, trade associations, sister agencies, and state and local authorities. Messages should be crafted and channels selected to separately address the perspectives of specific stakeholder groups. The governance structure for the project is key to the "by whom to whom" question.

When should I communicate? Communication needs to start early and occur frequently, not only to convey specific information about a project, but also to establish the integrity and credibility of the project champion and manager. "Putting it out there" early and often—rather than chasing an audience after other sources have generated information—is one way to demonstrate leadership and gain the respect of the audience.

Beyond the general guidance to communicate early and often to demonstrate openness and establish credibility, we recommend communicating during the planning phase of a project to define the project objectives and to identify risks and mitigation strategies. Continuing communication during the monitoring and measuring phase of a project is equally important, to share information needed for decision making or to garner support for decisions, to manage risks and impacts, and to maintain an iterative process of measuring and analyzing progress.

How do I get my message out? Effective communication uses specific channels to target specific audiences. Channels—the conduits for sending your message to the chosen audience—can take many forms, including web-based knowledge centers, formal written reports, e-mail, websites, workshops, seminars, newsletters, written letters, and the like. How formal or informal the channel should be depends on factors such as the message itself, the audience for the message, and how long the information will be needed.

Questions to consider when selecting a channel include: Where or from whom does a particular audience sector get its information? Where does this audience spend most of its time? and Where are they most likely to give you their attention?

One communications channel that is gaining in popularity and offers many benefits is the web-based knowledge center. Web-based knowledge centers enable entire organizations, indeed multiple organizations, to capture, reuse, update, and reorganize material at a single easy-to-maintain site. One such example is Acquisition Solutions' Virtual Acquisition Office One-Stop™ sites. Such web portals can include internal directives, policies, and procedures; notices; reports; feedback surveys; topical literature and research; and interactive information exchange such as work areas, threaded messaging, and the like. In addition, these web centers can be easily shared or cosponsored with communication "partners." An often overlooked facet of "how" to communicate is the potential to establish partners to help get out messages. For example, procurement offices can partner with program offices or—shocking concept though it might be—even with oversight organizations, to communicate information to generate desired behaviors for an organizational transformation or for undertaking a results-based approach to an acquisition.

ARE THERE ANY SPECIAL FACTORS THAT COULD HELP IN COMPOSING AN EFFECTIVE MESSAGE?

The National Institutes of Health (NIH) is an organization that has honed the art of communication over the years. We can apply its experience with health programs to achieve results in federal acquisition. The following factors are based on NIH guidance for constructing messages to achieve the desired outcome:[2]

■ *Clarity* – Messages should convey information clearly, to ensure the recipients' understanding and to limit the chances for inappropriate action. Avoid technical and bureaucratic terms as much as possible and eliminate information the recipient does not need to make decisions (such as unnecessarily detailed explanations).

- *Consistency* – In an ideal world, there would be consensus on the meaning of information, and all messages would be received in the intended way. However, different audiences, and even different members within an audience, may interpret data differently, a factor to consider when constructing the message.
- *Main points* – The main points should be stressed, repeated, and never hidden within less strategically important information.
- *Tone and appeal* – A message should be reassuring, alarming, challenging, or straightforward, depending on the desired impact and the target audience. Messages should always be truthful and honest.
- *Credibility* – The source of the message and information should be believable and trustworthy.
- *Recipient need* – For a message to break through the "information clutter," it should be based on what the target audience perceives as important, what it wants to know, and not what is most important or most interesting to the message originator.

Prior to final production, it's a good idea to test the message on a sample audience to ensure understanding and the intended reaction.

What information should a project communications plan cover?

Project communications plans, should address the following:
- Purpose
- Communications approach
- Marketing approach
- Cultural transformation approach
- Organizational information needs
- Stakeholders information needs.

See www.spinproject.org for a strategic communications plan template licensed under a Create Commons license that permits free use of its contents for any noncommercial purpose, provided the SPIN project is credited.

Developing a preliminary project communications plan is among the first steps an organization should undertake in setting the stage for successful performance after contract award. When communicating to push for results, the project communications plan should address not only distribution of status and other data, but also ways to foster the desired culture. The project communications plan might, for example, include workshops, seminars, newsletters, websites, e-mail, or other means of communication that will help garner support for the pending contract and prepare people for the new performance-based business environment.

How can project managers ensure their communications projects help obtain project results?

Whatever the vehicle, communications planning should take into consideration the information needs and preferred communication styles of all the stakeholders. The project communication plan should identify who is receiving the information and what method will be used to distribute it. The messages in all cases should be proactive, honest, consistent, frequent, open, and transparent. Following is a list of effective project communications principles Acquisition Solutions developed for a client with stakeholders in the state and local communities.

Figure 7-2. Top 10 Principles of Effective Project Communications Checklist
❏ Honesty is the only policy.
❏ Segment the audience.
❏ Use multiple channels.
❏ Use multiple voices.
❏ Communicate clearly.
❏ Communicate proactively and frequently.
❏ Anticipate and act on barriers.
❏ Provide opportunities for feedback.
❏ Build a sense of "connectedness."
❏ Tailor messages to specific stakeholder groups.

1. *Honesty is the only policy.* The truth is almost always less terrible than what people imagine if they hear nothing. The truth buys credibility, and the mere fact of speaking the truth shows that something profoundly different is under way. Demonstrate an appropriate level of openness: If we know, we'll tell you. If we don't know, we'll tell you we don't know. If we can't tell you, we'll tell you why, and give you an estimated time when we can tell you.

2. *Segment the audience.* Acknowledge the fact that different areas have different needs with respect to communication. Key questions to consider include:

- Who are the key audiences?
- How will they be affected by the change?
- What reaction will they have to it?
- What behavior do we need of them?
- What messages do they need to hear for that behavior to be stimulated?
  - When do they need to hear these messages?
  - What medium should we use for each message?
  - Who should communicate the message to them?

3. *Use multiple channels.* Use a variety of vehicles to ensure maximum penetration and receptivity.

4. *Use multiple voices.* Not everyone will connect with a particular individual or identify with a given perspective. Use various agency leaders to ensure credibility.

5. *Communicate clearly.* The content of the message must be clear, specific, and comprehensible, noting as applicable the purpose, progress, process, and problems.

6. *Communicate proactively and frequently.* The key to effective communication is reinforcement: repeat your message in many ways, through many channels, and by many people. The "Rule of Seven" applies: the same thing must be communicated seven times in seven different ways before anybody will believe it. Proactive communication is better than reactive communication. Reactive communication typically is not well thought out and therefore is not strategic. Reactive communication tends to be written, formal, and delivered impersonally.

7. *Anticipate and act on barriers.* Refer to the audience assessment. Anticipating barriers and concerns can help you structure each communication to address those concerns.

8. *Provide opportunities for feedback.* Respond to all feedback and publicize resulting actions or decisions. Allow recipients of the communication to give feedback and provide mechanisms for them to do so.

9. *Build a sense of "connectedness."* Every stakeholder has specific needs; however, communication mechanisms can foster a sense

of connectedness by using the same general messages enterprise-wide and establishing a single communication base.

10. *Tailor messages to specific stakeholder groups.* In addition to enterprise-wide messages, tailor specific messages to the individual needs and roles of each stakeholder group. Create messages to fit the end-user profile. Keep messages clear and concise to enable best transmission, especially when communicating the future vision. Anticipate questions. Cascade messages from level-to-level to reach all audience groups, and use the leadership team to lead the effort.

To be effective and consistent with your message, communications principles should be engrained in all personnel associated with performance of the project.

HOW ARE GOVERNANCE BODIES INCORPORATED INTO A COMMUNICATIONS PLAN?

A good communications plan identifies how the governance bodies will accomplish status reviews and updates and how they will disseminate information to the organizations and stakeholders involved in the program. One approach is summarized in Figure 7.3.

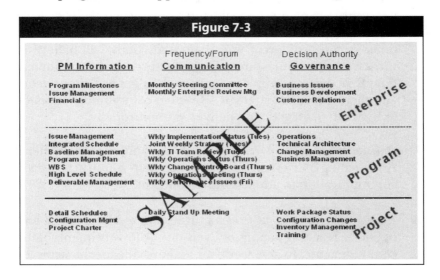

Figure 7-3

PM Information	Frequency/Forum Communication	Decision Authority Governance	
Program Milestones	Monthly Steering Committee	Business Issues	*Enterprise*
Issue Management	Monthly Enterprise Review Mtg	Business Development	
Financials		Customer Relations	
Issue Management	Wkly Implementation Status (Tues)	Operations	*Program*
Integrated Schedule	Joint Weekly Strategy (Tues)	Technical Architecture	
Baseline Management	Wkly TI Team Review (Tues)	Change Management	
Program Mgmt Plan	Wkly Operations Status (Thurs)	Business Management	
WBS	Wkly Change Control Board (Thurs)		
High Level Schedule	Wkly Operations Meeting (Thurs)		
Deliverable Management	Wkly Performance Issues (Fri)		
Detail Schedules	Daily Stand Up Meeting	Work Package Status	*Project*
Configuration Mgmt		Configuration Changes	
Project Charter		Inventory Management	
		Training	

WHAT IS THE DESIRED OUTPUT OF A PROJECT COMMUNICATIONS PLAN?

It is the ultimate goal of nearly every project manager to ensure three things:

- Provide clear, concise, and compelling project communications
- Deliver effective and efficient project communications
- Achieve high performance results that meet or exceed all customer and stakeholder expectations.

CONCLUSION

The ability to successfully communicate is a hallmark of good project leadership. Delivery of the right message at the right time in the right way is critical to building stakeholder buy-in and, therefore, to achieving project results. Developing and implementing a project communications strategy and plan are important steps often overlooked by managers undertaking projects or change management initiatives, but developing a project communications plan is not difficult. Tools are available to help develop and deliver the message; the principles, templates, samples, and examples in this chapter can help. Project leaders should avail themselves of all these tools and techniques and then incorporate them effectively into their projects to deliver high performance results.

In the next chapter, we discuss the Fifth Discipline of Performance-Based Project Management – the practice of risk management.

QUESTIONS TO CONSIDER

1. How effectively and efficiently does your organization manage project communications?

2. Does your organization err on the side of too much or too little communication?

3. What are the major obstacles to successful project communications?

4. Does your organization create and implement project communications plans?

Endnotes

[1]   The ArtsMarketing project of the Arts and Business Council of Americans for the Arts, *http://www.artsmarketing.org/marketingresources/files/KelloggStratPlanToolkit.pdf.*

[2]   *Making Health Communication Programs Work: A Planner's Guide,* Office of Cancer Communications, National Cancer Institute, National Institutes of Health (1992).

CHAPTER  8

FIFTH DISCIPLINE – OPPORTUNITY & RISK MANAGEMENT

By Gregory A. Garrett and Shaw Cohe

INTRODUCTION

The government's missions to provide for global defense, home-land security, and services to citizens challenge us with increasingly complex programs and projects to manage. These challenges involve varying degrees of risk, some to a high degree, all of which need to be managed accordingly. Said simply, the desired outcome for every project in both government and industry is to meet or exceed all contract and project performance requirements. But numerous Government Accountability Office (GAO) and inspectors general (IG) findings purport that program managers do a poor job of managing their programs, projects, and contracts–and particularly managing risk.

Managing a project within an organization is a challenge. Managing a multisector workforce around a complex performance-based project, meaning a project involving a customer, principal supplier (prime contractor), and numerous supply-chain partners (vendors and subcontractors) with a combination of high-technology products and services, as illustrated in Figure 8-1, is a much greater challenge. This chapter provides a summary of the opportunity and risk management process that should be considered when managing such complex projects, which involve multiple parties and multiple functions.

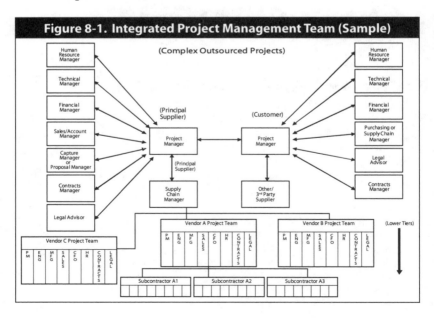

Figure 8-1. Integrated Project Management Team (Sample)

Managing opportunities and risks is an inherent part of everyone's life and is a crucial aspect of every project, but many people, including project managers, assume risks without ever formally assessing or attempting to mitigate them. Opportunity and risk management is an attempt to predict future outcomes based on current knowledge, so it is not a precise science. However, it is possible to increase opportunities and reduce risks or prevent risk events from occurring by using a process approach to opportunity and risk management.

Opportunity and risk management is an important element of both contract management and project management because every project contains elements of uncertainty, such as varying amounts of funding, changes in contract delivery date(s), changes in technical requirements, and increases or decreases in quantity. Opportunity and risk management should be thought of as a part of the performance-based project management methodology.

What is opportunity?

Opportunity is the measure of the probability of an opportunity event—a positive desired change—occurring and the desired impact of that event.

What is risk?

Risk is the measure of the probability of a risk event—an unwanted change—occurring and the associated effect of that event. In other words, risk consists of three components:

- A risk event (an unwanted change)
- The probability of occurrence (uncertainty)
- The significance of the impact (the amount at stake).

How do you define risk in the world of U.S. federal acquisition management?

Risk is defined by the U.S. Department of Defense (DoD) as a measure of future uncertainties in achieving program performance goals and objectives within defined cost, schedule, and performance constraints. Specific areas of potential risk are provided in Table 8-1.

Table 8-1. DoD: Risk Areas, Definitions & Examples		
Risk Area	**Definition**	**Significant Risks**
Threat	Sensitivity to uncertainty of threat description	Uncertainty in threat accuracy.
		Sensitivity of design and technology to threat.
		Vulnerability of system to threat and threat countermeasures.
		Vulnerability of program to intelligence penetration.
Requirements	Sensitivity to uncertainty in the system description and requirements	Operational requirements vaguely stated or not properly established.
		Requirements not stable.
		Required operating environment not described.
		Requirements do not address logistics and suitability.
		Requirements too constrictive—identify specific solutions that force high cost.
Design	Degree to which system design could change if the threat parameters change	Design implications not sufficiently considered in concept exploration.
		System will not satisfy user requirements.
		Mismatch of user manpower or skill profiles with system design solution or human machine interface problems.
		Increased skills or more training requirements identified late in the acquisition process.
		Design not cost effective.
		Design relies on immature technologies or "exotic" materials to achieve performance objectives.
		Software design, coding, and testing.
Test and Evaluation	Adequacy and capability of test and evaluation program to assess performance specifications and whether system is operationally effective, suitable, and interoperable	Test planning not initiated early in program.
		Testing does not address the ultimate operating environment.
		Test procedures do not address all major performance and suitability specifications.
		Test facilities not available to accomplish specific tests, especially system-level tests.
		Insufficient time to test thoroughly.
Modeling and Simulation (M&S)	Adequacy and capability of M&S to support all life-cycle phases	Same risks as contained in the significant risks for test and evaluation.
		M&S are not verified, validated, or accredited for the intended purpose.
		Program lacks proper tools and M&S capability to assess alternatives.

Table 8-1. DoD: Risk Areas, Definitions & Examples		
Risk Area	**Definition**	**Significant Risks**
Technology	Degree to which technology has demonstrated maturity to meet program objectives	Program depends on unproven technology for success—there are no alternatives.
		Program success depends on achieving advances in state-of-the-art technology.
		Potential advances in technology will result in less than optimally cost-effective system or make system components obsolete.
		Technology has not been demonstrated in required operating environment.
		Technology relies on complex hardware, software, or integration design.
Logistics	Ability of the system configuration and documentation to achieve logistics objectives	Inadequate supportability late in development or after fielding, resulting in need for engineering changes, increased costs, and/or schedule delays.
		Life-cycle costs not accurate because of poor logistics supportability analyses.
		Logistics analyses results not included in cost-performance trade-offs.
		Design trade studies do not include supportability considerations.
Production/ Facilities	Ability of the system configuration to achieve the program's production objectives	Production implications not considered during concept exploration.
		Production not sufficiently considered during design.
		Inadequate planning for long lead items and vendor support.
		Production processes not proven.
		Prime contractors do not have adequate plans for managing subcontractors.
		Sufficient facilities not readily available for cost-effective production.
		Contract offers no incentive to modernize facilities or reduce cost.
Concurrency	Sensitivity to uncertainty resulting from combining or overlapping phases or activities	Immature or unproven technologies will not be adequately developed before production.
		Production funding will be available too early—before development effort has sufficiently matured.
		Concurrency established without clear understanding of risks.

Table 8-1. DoD: Risk Areas, Definitions & Examples		
Risk Area	**Definition**	**Significant Risks**
Industrial Capabilities	Abilities, experience, resources, and knowledge of the provider to design, develop, manufacture, and support the system	Developer has limited experience in specific type of development.
		Contractor has poor track record relative to costs and schedule.
		Contractor experiences loss of key personnel.
		Prime contractor relies excessively on subcontractors for major development efforts.
		Contractor will require significant capitalization to meet program requirements.
Cost	Ability of system to achieve life-cycle support objectives; includes effects on budgets, affordability, and effects of errors in cost estimating techniques	Realistic cost objectives not established early.
		Marginal performance capabilities incorporated at excessive costs; satisfactory cost-performance trade-offs not done.
		Excessive life-cycle costs due to inadequate treatment of support requirements.
		Significant reliance on software.
		Funding profile does not match acquisition strategy.
		Funding profile not stable from budget cycle to budget cycle.
Schedule	Sufficiency of time allocated for performing the defined acquisition tasks	Schedule not considered in trade-off studies.
		Schedule does not reflect realistic acquisition planning.
		Acquisition program baseline schedule objectives not realistic and attainable.
		Resources not available to meet schedule.
Management	Degree to which program plans and strategies exist and are realistic and consistent	Acquisition strategy does not give adequate consideration to various essential elements, e.g., mission need, test and evaluation, technology, etc.
		Subordinate strategies and plans are not developed in a timely manner or based on the acquisition strategy.
		Proper mix (experience, skills, stability) of people not assigned to Program Management Office or to contractor team.
		Effective risk assessments not performed or results not understood and acted on.
External Factors	Availability of government resources external to the program office required to support the project, such as facilities, resources, personnel, government-furnished equipment, etc.	External government resources are unknown or uncertain.
		Little or no control over external resources.
		Changing external priorities are a threat to performance.

Table 8-1. DoD: Risk Areas, Definitions & Examples		
Risk Area	**Definition**	**Significant Risks**
Budget	Sensitivity of program to budget changes and reductions	Budget practices (releasing funds quarterly or monthly) negatively affect long-term planning processes.
		Budget changes or reductions can negate contractual arrangements and continuity of operations.
Earned Value Management (EVM) System	Adequacy of the contractors EVM process and realism of integrated baseline	Baseline proves unrealistic.
		Accurate and meaningful measures difficult to obtain.
		Contractor's EVM system does not effectively support project tracking.

Similarly, for the U.S. federal government, the Office of Management and Budget (OMB) identified in Circular A-11 19 different categories of risk to be addressed by government agencies in Exhibit 300 capital planning documentation. The OMB categories and the definitions used by the Department of Veterans Affairs are illustrated in Table 8-2.

Table 8-2. OMB: Risk Categories, Definitions & Examples		
Risk Category	**Definition**	**Examples**
Schedule	The risk that the project will not meet all or parts of its list of terminal elements with assigned start and finish dates, such as release(s), milestone(s), deliverable(s), or critical task(s).	(1) Project planning has the potential to become more complex than anticipated and could require significantly more time than estimated.
		(2) Implementation of the project is dependent on the completion of other projects; their delay would cause this project to be delayed.
Initial Cost	The risk that the quality of cost estimates and the ability to secure and manage budgetary resources for what is needed during the planning, preliminary engineering, and project design phases will be insufficient.	(1) Estimates are based on interdependent projects that cannot be analyzed as a single entity, and project design and engineering is one of those interdependencies.
		(2) The project will use new and relatively unproven technologies for which there are no comparable VA or federal examples to benchmark preliminary design or engineering costs.
Life-Cycle Costs	The risk that there will be insufficient funds to take the project through the overall process of developing an information technology (IT) system from investigation of initial requirements through analysis, design, implementation, and maintenance.	(1) Estimates are based on many iffy life-cycle cost assumptions (e.g., inflation rates).
		(2) Life-cycle costs can exceed estimates if the reliability of a system falls below projections.
		(3) Errors may exist in the cost-estimating technique used to derive the life-cycle costs.

Table 8-2. OMB: Risk Categories, Definitions & Examples

Risk Category	Definition	Examples
Technical Obsolescence	The risk that key technologies used in a project will lose value because a new, more functional product or technology has superseded the project's or when the project's product(s) becomes less useful or useless due to changes before the project has completed its full functional life cycle.	(1) Strategies for avoiding the use of outdated technical resources over the system life-cycle have not been incorporated into the project plan. (2) There is no plan for regular technology upgrades or refreshes.
Feasibility	The risk that a process, design, procedure, or plan can be successfully accomplished in the required time frame as proposed.	(1) The affected office(s) may not have the existing capability to successfully develop or implement the project within defined technical, scope, and schedule parameters. (2) There are no examples of successful implementation of the proposed approach, software, or hardware within public or private industry.
Reliability of Systems	The risk that the system, when operating under stated conditions, will perform its intended function acceptably for a specified period of time.	(1) The project may not have the ability to provide information about current or planned/desired system reliability. (2) The proposed system is new and there may be no commercial or government installations to benchmark for actual reliability data.
Dependencies and Interoperability between This System and Others	The risk that the project will fail because it depends on the successful completion of another system, or the project will not be able to work with other systems or products without an unplanned special effort.	(1) Reliance on interoperability with other systems (existing or in development) within the department and/or across the federal government (e.g., technical interfaces, schedule dependencies). (2) The project is directly linked to the long-term success, implementation, or ongoing maintenance of other systems.
Surety (Asset Protection) Considerations	The risk associated with the ability of the project to meet its obligations or when there is some public or private interest that requires protection from the consequences of a contractor's default or a project's delinquency.	(1) The loss or damage that may result from a contractor failing to deliver as promised. (2) The potential for substantial loss of capability of the project due to an unforeseen disaster and a lack of continuity of operations and/or disaster recovery plans.
Risk of Creating a Monopoly for Future Procurements	The risk associated with the use of closed or proprietary software/source code, as well as the dependence on a single vendor or product, which in turn creates a risk that in the future the contractor will be able to reap windfall profits by charging excessive costs or reducing service quality.	(1) The inability to conduct open competition in the future due to a current or planned procurement. (2) Use of non-open source code software. (3) The inability to connect with existing or planned department systems without extensive customization.

Table 8-2. OMB: Risk Categories, Definitions & Examples

Risk Category	Definition	Examples
Capability of Agency to Manage the Investment	Risk associated with an inexperienced project owner and management team, or the lack of established OMB-approved management tools, or performance indicators that show that the department cannot deliver the project as promised.	(1) Project manager(s) does not have a Project Management Professional or equivalent certification. (2) Department may lack an EVMS that meets OMB standards. (3) Earned value data show unexplained project cost and/or schedule variances greater than10 percent. (4) Contractors do not use earned value reporting.
Overall Risk of Project Failure	Risk that there is an inherent project weakness, such as the project missing a clear link between it and the organization's key strategic priorities, including agreed measures of success.	(1) Veterans and health providers who are the customers for the project's deliverables may desire a different solution. (2) Investment solution may be overtaken by activities being pursued outside the department (e.g., centralization of government functions into Centers of Excellence).
Organizational and Change Management	Risk that activities involved in: (1) defining and instilling new values, attitudes, norms, and behaviors within an organization that support new ways of doing work and overcome resistance to change; (2) building consensus among customers and stakeholders on specific changes designed to better meet their needs; and (3) planning, testing, and implementing all aspects of the transition from one organizational structure or business process to another will not be successful.	(1) Organizational and/or departmental cultural resistance to proposed process change may be high. (2) Extensive employee training may be required to apply benefits of investment to existing or proposed process. (3) Initial operation of the new system demonstrates lack of use, improper use, or failure to fully use due to unchanged organizational structure or process.
Business	Risk that an investment will fail to achieve the expectations of the project's owners and customers.	(1) Investment's statements of support of department customers not carried through in project outcomes. (2) Investment planning has little or no customer involvement.
Data/ Information	Risk associated with data/ information loss or disruptions caused by natural disasters (hurricanes, tornadoes, floods, earthquakes, etc.) or by area-wide disruptions of communication or electric power or malicious attacks; also can include the ability of the investment to obtain, store, produce, share, and manipulate data as planned.	(1) No contingency plans exist to deal with the loss/misuse of data or information. (2) Project may not be able access data from other sources (federal, state, and/or local agencies).

Table 8-2. OMB: Risk Categories, Definitions & Examples

Risk Category	Definition	Examples
Technology	This risk refers to the problems associated with the use of technologies new to the department, new software releases, or hardware new to the market.	(1) Immaturity of commercially available technology. (2) Technical problems/failures associated with applications to be used. (3) Inability to provide planned and desired technical functionality. (4) Possibility that the application of software engineering theories, principles, and techniques will fail to yield the appropriate software product. (5) Final product will be overly expensive, delivered late, or otherwise unacceptable to the customer.
Strategic	The risk of misalignment with department mission and strategic goals and/or the President's Management Agenda.	(1) The investment fails to achieve those strategic goals it states it will support. (2) Project objectives are not clearly linked to the department's overall strategies or to government-wide policies and standards.
Security	The risk that the investment does (or will) not conform to applicable department and/or federal security standards.	(1) The investment does not have a current security plan. (2) The systems associated with the investment do not have current certification and accreditations (C&As). (3) The project's contractors are not in full compliance with department or federal security requirements. (4) Security training is not at 100 percent compliance.
Privacy	The risk of possible violations of the legal restrictions on the collection, use, maintenance, and release of information about individuals.	(1) Investment may feature a publicly accessible web site with personal data links. (2) Investment may involve a process that collects, manipulates, stores, or shares personally identifiable information. (3) Investment may convert paper files with personal data to electronic files.
Project Resources	The risk that assets available or anticipated, including people, equipment, facilities, and other things used to plan, implement, and maintain your project will be insufficient.	(1) The scope of the investment is not clear. (2) Necessary project resources are not clearly or completely specified. (3) No examples of a successful approach to solving the problem are provided in either project description or in discussion of alternatives.

WHAT IS OPPORTUNITY AND RISK MANAGEMENT?

The primary goal of opportunity and risk management (ORM) is to continually seek ways to maximize opportunities and mitigate risks. ORM is an iterative process approach to managing those opportunities and risks that may occur during the course of business that could affect the success or failure of the project. Once identified, the probability of each event's occurrence and its potential effect on the project are analyzed and prioritized, or ranked from highest to lowest. Beginning with the highest prioritized events and working down, the project team determines what options or strategies are available and chooses the best strategy to maximize opportunities and to reduce or prevent the identified risks from occurring. This information is the basis for the ORM plan, which should be continually referred to and updated during the project life cycle.

HOW CAN ORM BE INTEGRATED INTO PROJECT MANAGEMENT?

Some business managers rely solely on their intuitive reasoning (ability to guess correctly) as the basis for their decision making. But in today's complex systems environment, an astute business manager understands the importance of using a highly skilled project team to identify both opportunity and risk events, assess possible effects, and develop appropriate strategies to increase opportunities and reduce risks. A project work breakdown structure (WBS) is an effective means of relating project tasks to possible opportunities and risks.

To integrate ORM into project management successfully, the project manager must ensure that an ORM plan is included as part of the overall business management planning process. It is vital that ORM become a mind-set for all business professionals, especially project managers and contract managers, both government and contractor.

Figure 8-2 lists the key inputs, numerous proven tools and techniques to increase business opportunities and mitigate project risks, and the desired outputs that should be considered when managing complex performance-based projects.

Figure 8-2. The Opportunity & Risk Management Process

Key Inputs	Tools & Techniques	Desired Outputs
People	ORM Model	Maximize Opportunities
Elements of Opportunity	Idea Generation and Profitability	Mitigate Risks
Elements of Risk	Measurement Summaries	Deliver Successful Projects
Corporate Culture (Risk-Taking vs. Risk-Averse)	Project Complexity Assessment Tool	
Training	Checklist of Software Risks	
	Software Engineering Tables	
	Project Risk Management Plan Outline	
	Project Risk Mitigation Form	
	Types of Contracts — Risk Sharing Tools	
	Project Doability Analysis Form	
	Project Bid/No Bid Assessment Tool	
	Project Phases and Control Gates	
	ORM Decision — Support Software Matrix	

Inputs

The following are all key inputs to the ORM process, which should be used when managing complex performance-based projects.

- **People** – Given that multiple parties, usually consisting of people from numerous sectors and functional disciplines (project management, engineering, contracts, finance, manufacturing, purchasing/supply-chain management, quality, etc.), typically are involved with complex performance-based projects, as illustrated in Figure 8-1, the need for effective opportunity and risk management is great.

- **Elements of Opportunity**[1]
  - *Strategic Alignment* refers to how consistent the project opportunity is with the core mission or business or corporate direction for new business. Companies have a much higher probability of winning and being successful

during delivery when the project opportunity is consistent with their core business and strategic direction. Agencies have a much higher probability of achieving mission success through strategic alignment of mission, project, and contract objectives.

- *Competitive Environment* refers to whether your company or your competitor is perceived by the customer as the product/service/solution leader and, thus, favored as the key supplier. Opportunities when the customer perceives your company as the leader and the favored supplier (for reasons other than price) are highly desirable. Customers may have this perception due to technology, reputation, past experience, industry commitment, and so on. For federal teams, a performance-based competition can be managed to promote a competition of ideas and price to support the mission and achieve the objectives.

- *Project Value* refers to the dollar value of the project. The intent is to distinguish "small" from "large" revenue or value opportunities. Obviously, this needs to be assessed in the context of the size of your company or in relation to your agency's mission, budget, and rank-ordering of projects.

- *Expected Margin* refers to the likely margins on the business given the competitive environment and what it will take competitors to win.

- *Future Business Potential* refers to degree to which this project will affect additional business beyond the scope of the specific opportunity. For example, the opportunity may be a means to win a new project. Consider the degree to which specific identifiable future business is dependent on winning and successfully delivering this business.

- *Probability of Success* refers to the likelihood that the project will succeed and, for competing contractors, that they will win the business versus one of their competitors.

- *Collateral Benefit* refers to the degree to which pursuit of this project will improve existing skill levels or develop new skills that will benefit other projects or future business.

- *Project Importance* refers to the overall need to deliver mission and program objectives, as assessed by the project manager—or to win the project, as assessed by the sales manager or key account manager. This should be based on consideration of all the opportunity elements, along with any other tangible or intangible aspects of the opportunity that are considered relevant.

- ***Elements of Risk***[2]
  - *Customer Commitment* refers to the degree to which the customer has demonstrated a solid commitment to implement the products/services/solution offered in the project. Typically, this type of commitment is demonstrated through either budgeting for the implementation in a current or future business plan or identifying and assigning resources to support the implementation.
  - *Corporate Competence* refers to the company's past experience or core competencies to deliver the products/services/solution required in the project. The more past experience the company has in projects exactly like the project at hand, the lower the risk. Conversely, if the type of project has never been completed successfully by any company in the past, then there is high risk.
  - *External Obstacles* refers to the existence of roadblocks that are beyond the control of either the agency or the contractor. A good example of this would be if your customer were a regulated utility that must obtain approval from a state or federal authority before it can implement the project. Another example might be if the customer has yet to secure the budget needed to fund the implementation during a period when capital is tightly constrained.
  - *Opportunity Engagement* refers to the degree to which the contractor versus the contractor's competitors were involved or were likely to have been involved or may have been involved in establishing the customer's requirements. Some contractors believe that if they did not help the customer develop—or inform the development of—its requirements, chances are one of their competitors did. While this is not necessarily the case, government teams should keep in mind that good acquisition planning and market research inform the agency in appropriate ways prior to the development of the requirement, often from multiple contractors' perspectives.
  - *Solution Life-Cycle Match* refers to the degree to which the solution involves the use of existing mature products versus new products or leading-edge technology. If the solution involves mature products available today, the risk of the solution not working is very low. On the other hand, if the solution involves many new products that have yet to be released or are based on leading-edge technology, there is

a risk of encountering development delays or the products not working as planned.

- *Period of Performance* refers to the length of the project. The longer the project the greater the chance of significant changes. Personnel, customer environment, and business climate are a few examples of changes that can introduce risk affecting the project.
- *Delivery Schedule* refers to when delivery is required and who controls the schedule. From the contractor's viewpoint, the ideal situation is if the schedule is flexible and can be set by the company, which can then ensure adequate time to be successful. Conversely, if the government already has fixed the delivery schedule and also has identified penalties for missing schedules, the company will be assuming a risk associated with missing deliveries and may propose a higher price to mitigate risk.
- *Resource Coordination* refers to the number of internal and external groups that must be engaged to deliver the solution. The larger the number of internal groups required, the more coordination it will take to ensure successful delivery and the higher the risk of a disconnect and delivery problem. Coordination of outside suppliers or support groups typically introduces even more risk, as there generally is more control over internal groups than external suppliers to resolve problems.
- *Nonperformance Penalties* refers to the degree to which there are specified contract penalties for failure to deliver as promised. From the contractor's viewpoint, risk is mitigated if the customer has not specified penalties or there is the opportunity to negotiate them. If the customer has specified monetary or other penalties that are nonnegotiable, this increases risk and may be mitigated by higher prices.
- *Overall Feasibility* refers to the degree of feasibility of the project as assessed by a knowledgeable representative of the group accountable to deliver the solution. A major factor to consider in assessing feasibility is past experience in fulfilling obligations or addressing unforeseen problems equitably. If the project is extremely complex and the customer has a poor track record of supporting complex projects, there is a high risk that the project will not be implemented successfully.

- ***Corporate Culture*** – Refers to the tendency a organization has to either promote innovation and risk taking or to promote status quo and risk aversion.

- ***Training*** – Refers to the need for all individuals involved with managing complex performance-based projects to receive competency-based training on the ORM process and numerous related tools and techniques, discussed in this chapter.

Tools and Techniques

The following are a few of the many proven-effective tools and techniques to help business professionals involved in complex performance-based projects maximize the elements of opportunity and minimize or mitigate the elements of risk.

The Opportunity & Risk Management Model

The ORM Model (Figure 8-3) is an ongoing process model that has two major pieces: opportunity/risk assessment and opportunity/risk action plans. Opportunity/risk assessment is composed of three steps: identify opportunities and risks, analyze them, and prioritize them. Opportunity/risk action plans also are composed of three steps: develop opportunity and risk action plans and strategies, implement opportunity and risk action plans into the project management plan, and evaluate project results. Figure 8-3 illustrates the suggested six-step ORM model.

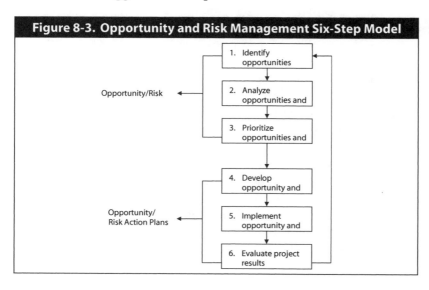

Figure 8-3. Opportunity and Risk Management Six-Step Model

Idea Generation & Profitability Measurement Summaries

Forms 8-1 and 8-2 provide a brief summary of some of the numerous tools and techniques available to gather information and analyze profitability, which can be valuable to the ORM process. Each of the following idea generation tools and techniques and profitability measurement tools and techniques is listed with a brief analysis of its respective advantages and disadvantages to support ORM in a performance-based project environment.

Form 8-1. Idea Generation (Opportunity and Risk Management) Tools and Techniques Summary		
Tool	**Advantages**	**Disadvantages**
Interview techniques	Can be used for many different specific applications Very flexible Gets access to knowledge project manager does not have	Is not a substitute for project manager decision making
Brainstorming	Interactive nature encourages group synergy	Stronger moderator required to prevent evaluation before end of idea generation Outspoken individuals can dominate the process Potential for "group think" Time intensive
Nominal group technique	Avoids inhibition problems associated with interactive methods Avoids tendency to start evaluation too early Avoids tendency to focus on single train of thought	Lack of multiperson synergy
Crawford slip	Same as nominal group Most efficient method in terms of number of ideas per work hour Complete paper trail of process	Same as nominal group Moderator must integrate large volume of data
Analogy techniques	Essential to learn from past mistakes Prevents "reinventing the wheel" If lessons-learned summaries exist, it can be easy to get data Can be used effectively in conjunction with interviewing techniques	Can be time intensive Relevant data may be difficult to obtain Data may not be accurate for new applications
Delphi technique	Avoids tendency to focus on single train of thought Allows "unbiased" expert judgment	Requires a lot of analysis Time consuming No direct group interaction
Affinity diagram	Expedites information gathering Eliminates dominant personalities from directing group decisions	No personal interaction No synergy

Form 8-2. Profitability Measurement (ORM) Tools and Techniques Summary		
Tool	**Advantages**	**Disadvantages**
Return on sales	Simplicity of use Simplicity of understanding Flexibility	Does not evaluate relative levels of investment
Return on assets	Reflects actual level of investment in effort Reflects time value of investment Easy method is relatively accurate	Does not reflect time value of return More complex than return on sales measurement
Internal rate of return	Most sensitive profitability measure Reflects timing of all inflows and outflows Reflects overall level of investment Very effective for sensitivity	Highly sophisticated and complex; not well understood Requires detailed knowledge of cash flow profile Requires use of automated tool

Project Complexity Assessment Tool

The Project Complexity Assessment Tool (PCAT) is a proven-effective means of assessing ten key project-related factors to determine the relative difficulty of managing a project. Once the project complexity has been assessed, an organization then can determine the appropriate level of project management support to ensure project success.

Project Complexity Assessment Tool

Overview

The Project Complexity Assessment Tool (PCAT) was designed in 1992 by the Center for Project Studies as a part of Garrett Consulting Services, to help an organization's Project Management Office (PMO) assess project difficulty. PCAT entails a two-step process: (1) Project Complexity Level Assessment, and (2) Project Complexity Continuum Charting (Form 8-3).

Project Complexity Level Assessment

By definition every project is somewhat unique, thus different projects require different levels of support to ensure success. The level of project management support required in a performance-based project is based on numerous items, including the following ten key assessment factors:

Project Value — Total estimated cost for successfully completing a performance-based project.

Project Profitability — Total estimated gross margin or profitability a principal supplier expects to achieve for successfully completing the project.

Professional Services — Percentage of the total financial value for services.

Project Technical Difficulty — Complexity of the hardware, software, integration, and overall project performance.

Project Risk — Evaluation of potential risk events, their potential impact, and their probability of occurrence.

Project Duration — Expected length of the project.

Extent of Outsourcing — Expected number of parties or companies directly under contract to the principal supplier and/or to the government for performance of the project.

Technological Maturity — Assessment of overall maturity of the hardware: proven, relatively new, or developmental.

Project Research and Development — Significance of research and development within the scope of the project.

Project Importance — Assessment in terms of mission criticality, market positioning, account relationship, or overall financial impact.

After an opportunity is identified as a project, the PMO, both government and contractor, should determine the appropriate level of project management support, using the following procedures.

Project Complexity Level Assessment Tool (Level Assessment)

Consolidate known customer requirements documentation.

Use the Project Complexity Level Assessment as a guideline to determine the project complexity, which drives the appropriate level of project management support required to achieve project success.

Use/complete the Project Complexity Continuum to assess the level of project complexity and appropriate project management staffing. A Level 3 complexity project requires the least project management support, while a Level 1 complexity project requires the highest or most project management support.

Project Complexity Level Assessment		
Project Complexity Level	**Project Assessment Factors**	**Assessment Criteria**
Level 3	Project Value	Less than $5 million (U.S.)
	Project Gross Margin	Less than 20% (margin to revenue)
	Project Professional Services	Low to moderate
	Project Technical Difficulty	Low or moderate
	Project Risk	Low to moderate
	Project Duration	12 months or less
	Extent of Outsourcing	Minimal
	Technological Maturity	Standard or nondevelopmental technology
	Project Research and Development	Minimal
	Project Importance	
		Minimal to moderate

Project Complexity Level Assessment		
Project Complexity Level	**Project Assessment Factors**	**Assessment Criteria**
Level 2	Project Value	$5 million to $50 million (U.S.)
	Project Gross Margin	20% to 40% (% margin to revenue)
	Project Professional Services	High, with multiple providers
	Project Technical Difficulty	Moderate to high
	Project Risk	High to very high
	Project Duration	Longer than 12 months
	Extent of Outsourcing	Multiple subcontractors or requirement to act as prime integrator
		Substantial developmental and/or Integration technology
	Technological Maturity	Significant software development
	Project Research and Development	
	Project Importance	High

Project Complexity Level Assessment		
Project Complexity Level	**Project Assessment Factors**	**Assessment Criteria**
Level 1	Project Value	Greater than $50 million (U.S.)
	Project Gross Margin	Greater than 40%
	Project Professional Services	High, with multiple providers
	Project Technical Difficulty	Very high
	Project Risk	Very high
	Project Duration	Multiple years
	Extent of Outsourcing	Multiple subcontractors or requirement to act as prime integrator
	Technological Maturity	Substantial developmental and/or integration technology
		Significant software development
	Project Research and Development	
	Project Importance	Very high

Project Complexity Assessment Tool
(Project Complexity Continuum Procedures)

As an additional means of determining the level of project complexity a project complexity continuum has been developed.

The Project Complexity Continuum consists of a number of questions concerning the project with a range of possible answers underneath. The PMO should mark the point on the continuum that best describes the factors of the particular project. Level 3 projects typically fall in the low to low-to-medium range for each of the ten assessment factors. Level 2 projects typically range from the low-to-medium to medium-to-high range for each assessment factor. Level 1 projects typically range from medium to high for each assessment factor. The bars associated with the aggregate weighting section in the Project Complexity Level Assessment provide a visual depiction of the differences between the levels.

After answering all of the questions, the PMO should determine the level most appropriate for the project to be managed. There is a core of essential project management activities that should be performed for each project.

Form 8-3. Project Complexity Assessment Tool (Project Complexity Continuum)

Determine where on the continuum each Project Assessment Factor falls and determine the appropriate level of project management support to achieve success in a cost-effective manner.

Aggregate Weighting:

Level 3 Projects					
Level 2 Projects					
Level 1 Projects					

Project Assessment Factor	Low	Low to Medium	Medium	Medium to High	High
Project Value (Total revenue in $millions)	Under $5	$5 to $10	$10 to $50	$50 to $100	Over $100
Project Gross Margin (Total margin in % of revenue)	Under 10%	10% - 15%	16% - 25%	26% - 40%	Over 40%
Project Professional Services (% of revenue)	10% or less	15% to 25%	25% to 35%	35% to 45%	Greater than 45%
Project Technical Complexity	Very Low	Low	Moderate	High	Very High
Technological Maturity	Very High	High	Moderate	Low	Very Low
Project Risk (Scope, schedule, cost, contract terms, and supplier factors)	Very Low	Low	Moderate	High	Very High
Project Duration	6 Months or Less	6 Months - 1 year	1 Year - 18 Months	18 Months - 2 Years	2 Years +
Extent of Outsourcing (% of labor cost)	None/Under 10%	Few/Under 0%	Multiple/under 50%	Few/Over 50%	Multiple/over 50%
Project Research and Development	None	Minimal R&D	30% or less	30% to 50%	Over 50%
Project Importance to Principal Supplier	Very Low	Low	Moderate	High	Very High

This project has been determined to be executed at level _____, and will need the project management resources, tools, and techniques required for that project level.

Checklist of Software Risks

Form 8-4 provides a checklist of common software risks that should be considered in any project involving software development.

Form 8-4. Checklist of Software Risks
❑ Overambitious schedule
❑ Overambitious performance
❑ Underambitious budget
❑ Overambitious/unrealistic expectations
❑ Undefined/misunderstood contract obligations
❑ Unfamiliar/untried/new technology or processes
❑ Inadequate software sizing estimate
❑ Unsuitable/lack of development process model
❑ Unfamiliar/untried/new hardware
❑ Inconsistent/underdefined/overdefined requirements
❑ Inadequately trained/inexperienced personnel
❑ Continuous requirement changes
❑ Inadequately trained/inexperienced management
❑ Inadequate software development plan
❑ Unsuitable organizational structure
❑ Overambitious reliability requirement
❑ Unsuitable/lack of software engineering methods/techniques
❑ Lack of adequate automation support
❑ Lack of political support/user need for project
❑ Inadequate risk analysis or management

From: *Software Engineering Risk Analysis and Management,* by Robert N. Charette (McGraw Hill, 1994).

Software Engineering Risk Tables

The following five tables were developed by the U.S. Air Force Systems Command, now referred to as Air Force Materiel Command (AFMC), to help manage risk on its software development projects. The tables are from the government publication "Risk Management Concepts and Guidance," by the Defense Systems Management College, 1993. While this publication is many years old, it contains useful/practical information to help identify and assess risk factors involved in performance-based projects involving software development.

USAF - Software Engineering Risk Tables

Table 8-3. Quantification of Probability and Impact of Technical Failure

TECHNICAL DRIVERS	MAGNITUDE		
	LOW (0.0 - 0.3)	MEDIUM (0.4 - 0.5)	HIGH (0.6 - 1.0)
REQUIREMENTS			
Complexity	Simple or easily allocated	Moderate, can be allocated	Significant or difficult to allocate
Size	Small or easily broken down into work units	Medium or can be broken down into work units	Large or cannot be broken down into work units
Stability	Little or no change to established baseline	Some change in baseline expected	Rapidly changing or no baseline
PDSS*	Agreed to support concept	Roles and missions issues resolved	No support concept or major unresolved issues
Reliability & Maintainability	Allocated to hardware and software components	Requirements can be defined	Can only be addressed at the total system level
CONSTRAINTS			
Computer Resources	Mature, growth capacity within design, flexible	Available, some growth capacity	New development, no growth capacity, inflexible
Personnel	Available in place, experienced, stable	Available but not in place, some experienced	High turnover, little or no experience, not available
Standards	Appropriately tailored for application	Some tailoring, all not reviewed for applicability	No tailoring, none append to the contract
Government Furnished Property	Meets requirements, available	May meet requirements, uncertain availability	Not compatible with system requirements
Environment	Little or no impact on design	Some impact on design	Major impact on design
TECHNOLOGY			
Language	Mature, approved HOL** used	Approved or non-approved HOL	Significant use of assembly language
Hardware	Mature, available	Some development or available	Total new development
Tools	Documented, validated, in place	Available validated, some development	Unvalidated, proprietary, major development
Data Rights	Fully compatible with support and follow-on	Minor incompatibilities with support and follow-on	Incompatible with support and follow-on
Experience	Greater than 3 to 5 years	Less than 3 to 5 years	Little or none
DEVELOPMENTAL APPROACH			
Prototypes & Reuse	Used, documented sufficiently for use	Some use and documentation	No use and/or no documentation
Documentation	Correct and available	Some deficiencies, available	Nonexistent
Environment	In place, validated, experience with use	Minor modifications, tools available	Major development effort
Management Approach	Existing product and process controls	Product & process controls need enhancement	Weak or nonexistent
Integration	Internal and external controls in place	Internal or external controls not in place	Weak or nonexistent
IMPACT	Minimal to small reduction in technical performance	Some reduction in technical performance	Significant degradation to nonachievement of technical performance

From: Risk Management Concepts and Guidance, Defense Systems Management College, 1993
* Postdeployment support software
** Higher order language

USAF - Software Engineering Risk Tables

Table 8-4. Quantification of Probability and Impact of Schedule Failure			
	MAGNITUDE		
SCHEDULE DRIVERS	**LOW** **(0.0 - 0.3)**	**MEDIUM** **(0.4 - 0.5)**	**HIGH** **(0.6 - 1.0)**
RESOURCES			
Personnel	Good discipline mix in place	Some disciplines not available	Questionable mix and/or availability
Facilities	Existent, little or no modification	Existent, some modi-fication	Nonexistent, extensive changes
Financial	Sufficient budget allocated	Some questionable allocations	Budget allocation in doubt
NEED DATES			
Threat	Verified projections	Some unstable aspects	Rapidly changing
Economic	Stable commitments	Some uncertain commit-ments	Unstable, fluctuating commitments
Political	Little projected sensitivity	Some limited sensitivity	Extreme sensitivity
Government-Furnished Property	Available, certified	Certification or delivery questions	Unavailable and/or uncertified
Tools	In place, available	Some deliveries in question	Uncertain delivery dates
TECHNOLOGY			
Availability	In place	Some aspects still in development	Totally still in development
Maturity	Application verified	Some applications verified	No application evidence
Experience	Extensive application	Some application	Little or none
REQUIREMENTS			
Definition	Known, baseline	Baseline, some unknowns	Unknown, no baseline
Stability	Little or no change projected	Controlled change projected	Rapid or uncontrolled change
Complexity	Compatible with existing technology	Some dependency on new technology	Incompatible with exist-ing technology
IMPACT	Realistic, achievable schedule	Possible slippage in IOC*	Unachievable IOC
* Initial operational capacity			

USAF - Software Engineering Risk Tables

Table 8-5. Quantification of Probability and Impact of Cost Failure			
COST DRIVERS	**MAGNITUDE**		
	LOW **(0.0 - 0.3)**	**MEDIUM** **(0.4 - 0.5)**	**HIGH** **(0.6 - 1.0)**
REQUIREMENTS			
Size	Small, noncomplex, or easily decomposed	Medium, moderate complexity, decomposable	Large, highly complex, not decomposable
Resource Constraints	Little or no hardware-imposed constraints	Some hardware-imposed constraints	Significant hardware-imposed constraints
Application	Non-real-time, little system interdependency	Embedded, some system interdependency	Real-time, embedded, strong interdependency
Technology	Mature, existent, in-house experience	Existent, some in-house experience	New or new application. Little experience
Requirements Stability	Little or no change to established baseline	Some change in baseline expected	Rapidly changing or no baseline
PERSONNEL			
Availability	In place, little turnover expected	Available, some turnover expected	High turnover, not available
Mix	Good mix of software disciplines	Some disciplines inappropriately represented	Some disciplines not represented
Experience	High experience ratio	Average experience ratio	Low experience ratio
Management Environment	Strong management approach	Good personnel management approach	Weak personnel management approach
REUSABLE SOFTWARE			
Availability	Compatible with need dates	Delivery dates in question	Incompatible with need dates
Modifications	Little or no change	Some change	Extensive changes
Language	Compatible with systems & PDSS* requirements	Partial compatibility with requirements	Incompatible with system or PDSS requirements
Rights	Compatible with PDSS & competition requirements	Partial compatibility with PDSS, some competition	Incompatible with PDSS concept, noncompetitive
Certification	Verified performance application compatible	Some application compatible test data available	Unverified, little test data available
TOOLS & ENVIRONMENT			
Facilities	Little or no modifications	Some modifications existent	Major modifications, nonexistent
Availability	In place, meets need dates	Some compatibility with need dates	Nonexistent, does not meet need dates
Rights	Compatible with PDSS & development plans	Partial compatibility with PDSS & development plans	Incompatible with PDSS & development plans
Configuration Management	Fully controlled	Some controls	No controls
IMPACT	Sufficient financial resources	Some shortage of financial resources, possible overrun	Significant financial shortages, budget overrun key
* Postdeployment support software			

USAF - Software Engineering Risk Tables

Table 8-6. Quantification of Probability and Impact of Operational Failure			
OPERATIONAL DRIVERS	**MAGNITUDE**		
	LOW (0.0 - 0.3)	**MEDIUM (0.4 - 0.5)**	**HIGH (0.6 - 1.0)**
USER PERSPECTIVE			
Requirements	Compatible with the user environment	Some incompatibilities	Major incompatibilities with "ops" concepts
Stability	Little or no change	Some controlled change	Uncontrolled change
Test Environment	Representative of the user environment	Some aspects are not representative	Major disconnects with user environment
Operational Test & Evaluation Results	Test errors/failures are correctable	Some errors/failures are not correctable before IOC*	Major corrections necessary
Quantification	Primarily objective	Some subjectivity	Primarily subjective
TECHNICAL PERFORMANCE			
Usability	User friendly	Mildly unfriendly	User unfriendly
Reliability	Predictable performance	Some aspects unpredictable	Unpredictable
Flexibility	Adaptable with threat	Some aspects are not adaptable	Critical functions not adaptable
Supportability	Responsive to updates	Response times inconsistent with need	Unresponsive
Integrity	Secure	Hidden linkages, controlled access	Insecure
PERFORMANCE ENVELOPE			
Adequacy	Full compatibility	Some limitations	Inadequate
Expandability	Easily expanded	Can be expanded	No expansion
Enhancements	Timely incorporation	Some lag	Major delays
Threat	Responsive to change	Cannot respond to some changes	Unresponsive
IMPACT	Full mission capability	Some limitations on mission performance	Severe performance limitations
* Initial operational capacity			

USAF - Software Engineering Risk Tables

Table 8-7. Quantification of Probability and Impact of Support Failure			
SUPPORT DRIVERS	**MAGNITUDE**		
	LOW **(0.0 - 0.3)**	**MEDIUM** **(0.4 - 0.5)**	**HIGH** **(0.6 - 1.0)**
DESIGN			
Complexity	Structure maintainable	Certain aspects difficult	Extremely difficult to maintain
Documentation	Adequate	Some deficiencies	Inadequate
Completeness	Little additional for PDSS incorporation	Some PDSS* incorporation	Extensive PDSS incorporation
Configuration Management	Sufficient, in place	Some shortfalls	Insufficient
Stability	Little or no change	Moderate, controlled change	Rapid or uncontrolled change
RESPONSIBILITIES			
Management	Defined, assigned responsibilities	Some roles and mission issues	Undefined or unassigned
Configuration Management	Single point control	Defined control points	Multiple control points
Technical Management	Consistent with operational needs	Some inconsistencies	Major inconsistencies
Change Implementation	Responsive to user needs	Acceptable delays	Nonresponsive to user needs
TOOLS & ENVIRONMENT			
Facilities	In place, little change	In place, some modification	Nonexistent or extensive change
Software Tools	Delivered, certified, sufficient	Some resolvable concerns	Not delivered, certified, or sufficient
Computer Hardware	Compatible with "ops" system	Minor incompatibilities	Major incompatibilities
Production	Sufficient for fielded units	Some capacity questions	Insufficient
Distribution	Controlled, responsive	Minor response concerns	Uncontrolled or nonresponsive
SUPPORTABILITY			
Changes	Within projections	Slight deviations	Major deviations
Operational Interfaces	Defined, controlled	Some "hidden" linkages	Extensive linkages
Personnel	In place, sufficient, experienced	Minor discipline mix concerns	Significant concerns
Release Cycle	Responsive to use requirements	Minor incompatibilities	Nonresponsive to user needs
Procedures	In place, adequate	Some concerns	Nonexistent or inadequate
IMPACT	Responsive software support	Minor delays in software modifications	Nonresponsive or unsupportable software
*Postdeployment support software			

Project Risk Management Plan Outline

Use the following outline to prepare a comprehensive project risk management plan.

Project Risk Management Plan Outline

1.0 Project Scope

Insert the scope statement or provide a brief summary of the project, including a description of the work to be accomplished, a description of the customer's goals and objectives for the project, a general description of how the project will be accomplished, and other pertinent information that will provide a good overview of the project.

2.0 Risk Event Descriptions

For each element of the work breakdown structure, identify any major risks involved in that element. Complete Risk Event Descriptions and Risk Event Results. Reference or include a copy of the WBS in the section. The process is carried out as follows:

2.1 Identify Risks. For each element of the WBS, identify any major risks associated with that element. Ensure that each risk event refers to a specific WBS element.

2.2 Analyze Risks and Calculate the Weighted Cost Impact. In analyzing the risks, make the assumption that the risk event identified will occur. Think in terms of the remedial activity that will need to take place to rectify the occurrence of the risk event.

Using the same guidelines used in calculating the cost of the WBS elements, calculate the rectification cost (impact) without any form of "padding" or risk adjustment. Analyze the risk event and apply a weighting to the impact on a scale of 1 through 5, as follows:

Weight 1: Has little potential to cause disruption of schedule, costs, or performance (quality). Increase the impact by 5%

Weight 2: May cause minor disruption of schedule, costs, or performance (quality). Increase the impact by 10%.

Weight 3: May cause some disruption of schedule, costs, or performance (quality). Increase the impact by 15%.

Weight 4: May cause major disruption of schedule, costs, or performance (quality). Increase the impact by 20%.

Weight 5: Could cause significant, serious disruption of schedule, costs, or performance (quality). Increase impact by 25%.

Finally, estimate the probability of the event occurring as a percentage (between 0.01 and 0.99), and calculate the weighted cost impact as follows:
• (Cost Impact + Cost Impact Increase) x Probability of Occurrence = Weighted Cost Impact
Example: For a risk event with an estimate cost impact of $2,750, a weight of 4, and probability of occurrence at 85%: ($2,750 + $550) x 0.85 = $2,805

2.3 Identify High-Risk Events. If a specific risk event has greater than 75% probability and/or the weighted cost impact is greater than 10% of the total project cost, the risk event is by definition high risk. For each high-risk event, create a separate and unique WBS element that identifies the work required and the weighted cost impact required to rectify the occurrence of the high-risk event. This WBS element must be flagged as a high-risk event element as distinct from a normal WBS element.

2.4 Develop Mitigation Strategies. Determine potential strategies for mitigating the risk either avoiding it, controlling it, or transferring it to another party. Assuming the risk consequences also is a potential mitigation strategy, but it is the least desirable. Evaluate the potential cost impact of the mitigation strategy and reflect that impact in the risk budget.

Project Risk Management Plan Outline (cont.)

2.5 <u>Establish the Risk Budget.</u> Each high-risk event will become a line item in the risk budget. The other risk events should be accumulated and used to establish the managerial reserve. An amount for contingency, or those events and circumstances not anticipated in any way, should be calculated based on experience. These amounts together become the risk budget portion of the Project Budget. The risk budget should be margined at the same rate as the Project Budget to establish the budget at selling price. This then is presented to the customer in the proposal as the project price.

3.0 Risk Reassessment Plan

Identify the major reassessment points for this project and ensure those reassessment points are identified in the project plan. At a minimum, high-risk events should be reassessed at the following times:

- Whenever major changes occur in the project or its environment
- Before major decision milestones
- Periodically, according to some predetermined schedule.

3.1 <u>Risk Management Timetable.</u> Indicate the timetable for risk management activities. Ensure that the key events are also reflected on the project schedule. Major milestones include:

- Completion of risk identification and analysis
- Risk prioritization
- Completion of mitigation strategy development
- Incorporation into project plans and WBS
- Key reassessment points
- Documentation of risk results.

Project Risk Mitigation Form

The following Form 8-5 provides a simple yet proven-effective means of documenting possible project risk events, planned risk mitigation strategies, actual project results, and appropriate follow-up actions, if needed.

Form 8-5. Project Risk Mitigation Form
Risk Event Results
Mitigation Strategy/Strategies Selected
Results of Mitigation Strategy/Strategies
Follow-up Actions Required (If Needed)

Contract-Related Tools & Techniques

Contract Types/Risk Sharing Tools

Table 8-8 provides a summary of the various different forms of contractual pricing arrangements often called contract types. The summary provides a discussion or listing of the essential elements, advantages, disadvantages, and suitability for the various contract types typically used in outsourcing products and services.

Table 8-8. Contract Types/Risk Sharing Tools			
Type	**Essential Elements and Advantages**	**Disadvantages**	**Suitability**
(FFP) Firm Fixed Price	Reasonably definite design or performance specifications available. Fair and reasonable price can be established at outset. Conditions for use include: Adequate competition Prior purchase experience of the same, or similar, supplies or services under competitive conditions Valid cost or pricing data Realistic estimates of proposed cost Possible uncertainties in performance can be identified and priced Sellers willing to accept contract at a level that causes them to take all financial risks Any other reasonable basis for pricing can be used to establish fair and reasonable price	Price not subject to adjustment regardless of seller performance costs. Places 100% of financial risk on the contractor. Places least amount of administrative burden on contract manager. Preferred over all other contract types. Used with advertised or negotiated procurements.	Commercial products and commercial services for which reasonable prices can be established
(FP/EPA) Fixed Price with Economic Price Adjustment	Unstable market or labor conditions during performance period and contingencies that would otherwise be included in contract price can be identified and made the subject of a separate price adjustment clause. Contingencies must be specifically defined in contract. Provides for upward adjustment (with ceiling) in contract price. May provide for downward adjustment of price if escalated element has potential of failing below contract limits. Three general types of EPAs, based on established prices, actual costs of labor or material, and cost indexes of labor or material.	Price can be adjusted on action of an industry-wide contingency that is beyond seller's control. Reduces seller's fixed-price risk. FP/EPA is preferred over any cost-reimbursable-type contract. If contingency manifests, contract administration burden increases. Used with negotiated procurements and, in limited applications, with formal advertising when determined to be feasible. Contract Manager must determine if FP/EPA is necessary either to protect seller and buyer against significant fluctuations in labor or material costs or to provide for contract price adjustment in case of changes in seller's established prices.	Commercial products and services for which reasonable prices can be established at time of award

Table 8-8. Contract Types/Risk Sharing Tools (cont)			
Type	**Essential Elements and Advantages**	**Disadvantages**	**Suitability**
(FPI) Fixed Price Incentive	Cost uncertainties exist, but there is potential for cost reduction or performance improvement by giving seller a degree of cost responsibility and a positive profit incentive. Profit is earned or lost based on relationship that contract's final negotiated cost bears to total target cost. Contract must contain target cost, target profit, ceiling price, and profit-sharing formula. Two forms of FPI: firm target (FPIF) and successive targets (FPIS). FPIF: Firm target cost, target profit, and profit-sharing formula negotiated into basic contract; profit adjusted at contract completion. FPIS: Initial cost and profit targets negotiated into contract, but final cost target (firm) cannot be negotiated until performance. Contains production point(s) at which either a firm target and final profit formula or a FFP contract can be negotiated. Elements that can be incentives: costs, performance, delivery, quality.	Requires adequate seller accounting system. Buyer must determine that FPI is least costly and award of any other type would be impractical. Buyer and seller administrative effort is more extensive than under other fixed-price contract types. Used only with competitive negotiated contracts. Billing prices must be established for interim payment.	Development and production of high-volume, multiyear contracts
Cost-Reimbursement Contracts (Greatest Risk on Buyer)			
Cost	Appropriate for research and development work, particularly with nonprofit educational institutions or other nonprofit organizations, and for facilities contracts. Allowable costs of contract performance are reimbursed, but no fee is paid.	Application limited due to no fee and by the fact that the buyer is not willing to reimburse seller fully if there is a commercial benefit for the seller. Only nonprofit institutions and organizations are willing (usually) to perform research for which there is no fee (or other tangible benefits).	Research and development; facilities
(CS) Cost Sharing	Used when buyer and seller agree to share costs in a research or development project having potential mutual benefits. Because of commercial benefits accruing to the seller, no fee is paid. Seller agrees to absorb a portion of the costs of performance in expectation of compensating benefits to seller's firm or organization. Such benefits might include an enhancement of the seller's capability and expertise or an improvement of its competitive position in the commercial market.	Care must be taken in negotiating cost-share rate so that the cost ratio is proportional to the potential benefit (that is, the party receiving the greatest potential benefit bears the greatest share of the costs).	Research and development that has potential benefits to both the buyer and the seller

Table 8-8. Contract Types/Risk Sharing Tools (cont)

Type	Essential Elements and Advantages	Disadvantages	Suitability
(CPIF) Cost Plus Incentive Fee	Research and Development is feasible and there is a high probability positive profit incentives for seller management can be achieved if costs are closely managed. Performance incentives must be clearly spelled out and objectively measurable. Fee range should be negotiated to give the seller an incentive over various ranges of cost performance. Fee is adjusted by a formula negotiated into the contract in accordance with the relationship that total cost bears to target cost. Contract must contain target cost, target fee, minimum and maximum fees, and fee adjustment formula. Fee adjustment is made at completion of contract.	Difficult to negotiate range between the maximum and minimum fees so as to provide an incentive over entire range. Performance must be objectively measurable. Costly to administer; seller must have an adequate accounting system. Used only with negotiated contracts. Appropriate buyer surveillance needed during performance to ensure effective methods and efficient cost controls are used.	Major systems development and other development programs in which it is determined that CPIF is desirable and administratively practical
(CPAF) Cost Plus Award Fee	Contract completion is feasible, incentives are desired, but performance is not susceptible to finite measurement. Provides for subjective evaluation of seller performance. Seller is evaluated at stated time(s) during performance period. Contract must contain clear and unambiguous evaluation criteria to determine award fee. Award fee is earned for excellence in performance, quality, timeliness, ingenuity, and cost-effectiveness and can be earned in whole or in part. Two separate fee pools can be established in contract: base fee and award fee. Award fee earned by seller is determined by the buyer and is often based on recommendations of an award fee evaluation board.	Buyer's determination of amount of award fee earned by the seller is not subject to disputes clause. CPAF cannot be used to avoid either CPIF or CPFF if either is feasible. Should not be used if the amount of money, period of performance, or expected benefits are insufficient to warrant additional administrative efforts. Very costly to administer. Seller must have an adequate accounting system. Used only with negotiated contracts.	Level-of-effort services that can only be subjectively measured and contracts for which work would have been accomplished under another contract type if performance objectives could have been expressed as definite milestones, targets, and measurable goals

	Table 8-8. Contract Types/Risk Sharing Tools (cont)		
Type	**Essential Elements and Advantages**	**Disadvantages**	**Suitability**
(CPFF) Cost Plus Fixed Fee	Level of effort is unknown, and seller's performance cannot be subjectively evaluated. Provides for payment of a fixed fee. Seller receives fixed fee regardless of the actual costs incurred during performance. Can be constructed in two ways: Completion form: Clearly defined task with a definite goal and specific end product. Buyer can order more work without an increase in fee if the contract estimated cost is increased. Term form: Scope of work described in general terms. Seller obligated only for a specific level of effort for stated period of time. Completion form is preferred over term form. Fee is expressed as percentage of estimated cost at time contract is awarded.	Seller has minimum incentive to control costs. Costly to administer. Seller must have an adequate accounting system. Seller assumes no financial risk.	Completion form: Advanced development or technical services contracts Term form: Research and exploratory development; used when the level of effort required is known and there is an inability to measure risk
(T&M) Time and Material	Not possible when placing contract to estimate extent or duration of the work, or anticipated cost, with any degree of confidence. Calls for provision of direct labor hours at specified hourly rate and materials at cost (or some other basis specified in contract). The fixed hourly rates include wages, overhead, general and administrative expenses, and profit. Material cost can include, if appropriate, material handling costs. Ceiling price established at time of award.	Used only after determination that no other type will serve purpose. Does not encourage effective cost control. Requires almost constant surveillance by buyer to ensure effective seller management. Ceiling price is required in contract.	Engineering and design services in conjunction with the production of suppliers, engineering design and manufacture, repair, maintenance, and overhaul work to be performed on an as-needed basis

Project Doability Analysis

A number of successful companies and organizations worldwide have found it useful to summarize all of their project-related opportunities and risks in one simple document, often just a few pages in length, called a Project Doability Analysis. Form 8-6 is a suggested Project Doability Analysis form, which can be tailored and used by any organization to briefly summarize the opportunities and risks before or after the organization moves forward or bids on said project. Of course, it usually is better to conduct a doability analysis before an organization begins or bids on a project. It is better to know what your organization must do to achieve project success, before your organization is contractually obligated to perform the work, especially on large, complex, performance-based projects.

Form 8-6. Project Doability Analysis	
Project Manager Doability Assessment: Yes ❑ No ❑	
Executive Summary	
Project Name:	
Customer:	
Location(s):	Estimated Revenue in US$:
Start Date:	Completion Date:
Prepared by:	Phone #:
Fax:	e-mail:
Describe the project requirements/deliverables:	
Evaluate the project technical requirements/availability/research & development (R&D):	
Evaluate the feasibility of the project schedule (attach milestone schedule):	
Evaluate the reasonableness of the project financial commitments (attach the project business case):	
Conduct high-level risk assessment. Consider the following risks if appropriate: pricing, payment terms, acceptance, warranty, liability, R&D, implementation, environmental, etc. (attach the risk management plan):	
Describe significant assumptions implicit in the evaluation of the technical, schedule, and financial commitments:	
Assess the skills of the selected project team members (experience, education, training, professional certifications, strengths, and weaknesses):	
Executive assessment of project: *Doable:* Yes ❑ No ❑	

Project Bid/No Bid Assessment Tool (PBAT)

In addition to the host of other tools and techniques discussed in this chapter, all used to help identify, analyze, prioritize, and/or strategize what actions should be taken to maximize project opportunities and/or mitigate risks, the Project Bid/No Bid Assessment Tool (PBAT) can help guide suppliers in their important bid/no bid decision-making process. Project bid/no bid decision making is a critical element of all complex performance-based projects. PBAT was designed more than 12 years ago by Garrett Consulting Services and has been tailored or customized for numerous organizations worldwide. PBAT is a proven-effective tool to help supplier organizations/companies quickly and effectively evaluate project opportunities and risks as a part of their business case development, thus supporting their bid/no bid decision making.

Project Phases & Control Gates

Every large, complex, performance-based project goes through various phases during its life cycle. Different organizations have given the distinct phases a wide variety of names or numerical designations, as illustrated in Figure 8-4. Typically, before a project transitions from one phase to another in the project life cycle, it must pass through a control gate. A control gate often is an executive approval process that evaluates project-related opportunities and risks and decides whether it is prudent to continue with said project. Project control gates are effective only if they accurately assess project status and truly weigh both opportunities and risks.

Unfortunately, in both the public and private sectors, projects often are delayed for extended periods because of overly bureaucratic executive reviews and approval processes—which too commonly are merely rubberstamps to continue project performance. If a control gate does not add value to a project, it should be eliminated. Project phases and control gates can be effective tools to evaluate project opportunities and risks throughout the project life cycle. The key is to use them only when appropriate; fewer phases and fewer control gates are better. Every project phase and each project control gate should be well understood, by all the parties involved in the project, add value to the process, and help maximize project success.

Figure 8-4. Project Phases & Control Gates								
← ————— Study Period ————— →			← —Implementation Period — →				← — Operations Period — →	
NASA								
Formulation			Implementation				Operations	
Pre-Phase A Preliminary Requirements Analysis	Phase A Mission Needs and Conceptual Trade Studies	Phase B Concept Definition	Phase C Design and Development		Phase D Fabrication, Integration, Test, and Certification		Phase E Pre-Operations	Phase F Operations/ Disposal
DoD								
Pre-Phase 0 Determination of Mission Need	Phase 0 Concept Exploration and Definition	Phase I Demonstration and Validation	Phase II Engineering and Manufacturing Development				Phase III Production and Deployment	Phase IV Operations and Support
Typical High-Tech Commercial Business								
Product Requirements Phase	Product Definition Phase	Product Proposal Phase	Product Devel. Phase	Engineer. Model Phase	Int. Test Phase	External Test Phase	Production Phase	Manufacturing, Sales, and Support Phase

Control Gates

Operational Approval	New Initiative Approval	System Concept Approval	Development Approval	Production Approval

Adapted from: *Visualizing Project Management,* by Forsberg, Mooz, Cotterman (John Wiley & Sons, Inc., 2001).

Output

By using the opportunity and risk management six-step process and the numerous related proven tools and techniques discussed in this chapter, it is possible to maximize opportunities, mitigate risks, and significantly improve project results. Clearly, the desired output of the ORM process and related tools and techniques is to achieve successful projects.

SUMMARY

This chapter provides a summary of the ORM process and related tools and techniques that should be considered when managing complex performance-based projects. Too many organizations do not take the time and effort to thoroughly identify, assess, and prioritize business opportunities and risks or to develop action plans and implement said plans to maximize business opportunities and mitigate risks.

Questions to Consider

1. How well does your organization manage risk?

2. Has your organization developed an opportunity and risk management (ORM) process that is consistently used on all of your major projects?

3. How does your organization evaluate project complexity? What factors are typically considered?

4. List below the proven tools and techniques your organization uses to maximize opportunities and mitigate risks.

5. Which type of contract does your organization use most commonly when outsourcing/buying products and/or services?

Endnotes

[1] Adapted from *The Capture Management Life-Cycle*, by Gregory A. Garrett and Reginald J. Kipke (CCH, 2003).

[2] Ibid.

CHAPTER 9

SIXTH DISCIPLINE — PERFORMANCE MANAGEMENT

By: Gregory A. Garrett

INTRODUCTION

So, what exactly is performance management? Said simply, it is a proactive approach to ensure that customers get what they want, when they want it, and the seller is able to make a fair and reasonable return on its investment. While this concept of performance management may sound simple, it can be very complex and difficult to achieve due to numerous factors, which are sometimes unique to a specific project.

Today, we live in a world of increasingly demanding customers who often want the latest and greatest solutions and technology delivered as fast as possible, on-time, on-budget, with flawless execution. Of course, this all sounds fabulous; however, reality often falls short of these expectations. In fact, recent studies of the Project Management Institute (PMI) and other organizations indicate that over 70 percent of information technology (IT) projects are delivered late, over budget, and do not achieve the customers' requirements.

This chapter is focused on the sixth and final discipline of performance-based project management (PBPM) – performance management – which is a holistic process-based approach of planning the work and working the plan. In this chapter, we discuss the key inputs to the performance management process approach. Plus, we review a number of valuable tools and techniques to more effectively plan and manage performance including use of a performance management plan (PMP), as well as balanced scorecard (BSC) methods, contract analysis, pre-performance conference, payment system, quality plan, earned value management system (EVMS), change control system, and dispute management system.

WHAT IS THE PERFORMANCE MANAGEMENT PROCESS APPROACH?

In performance-based project management, success is determined by how well a particular contract or project helps an organization achieve its overall strategic mission and goals. Performance Management ensures the project deliverables are properly planned, scheduled, budgeted, and baselined; data is analyzed; status assessed; and decisions made about any changes needed to achieve success. PBPM is a disciplined process approach which allows project managers to be proactive, minimize risk, and maximize positive outcomes.

In order to ensure customers get what they want when they want it, it really helps to have a process approach and a plan of action. The following diagram, Figure 9-1, illustrates the key inputs, tools and techniques, and desired outputs of our recommended Performance Management Process Approach.

Figure 9-1. Performance Management Process Approach

Key Inputs	Tools & Techniques	Desired Outputs
People	Performance Management Plan	On-Time Delivery
Contract	Balanced Scorecard Method	On-Budget
Past Performance/ Experience	Contract analysis and planning	On-Time Payment
Change requests	Pre-performance conference	Completion of work
Invoices and payments	Payment System	Meet or exceed all requirements
Contract administration policies	Quality Plan	High customer satisfaction
Project Budget	Earned Value Managment System (EVMS)	
Stakeholder expectations	Change control system	
	Dispute management system	

WHAT ARE THE KEY INPUTS?

Various elements of the project environment are vital to a successful performance management process. The key inputs to a successful approach consist of the following items:

- **People:** A project will only succeed with the right quality and quantity of people working together with a common vision, mission, values, shared goals, and sufficient resources to get the work done.
- **Contract:** The contract is the primary guide–the written agreement between the parties–for what the buyer and seller have each agreed to perform.
- **Past Performance and Experience:** Both the buyer and seller bring with them certain skills, expertise, and baggage based upon their past performance and experience.
- **Change Requests:** Change requests are a common element of most contracts. An effective process for managing change must be in place to ensure that all requests are handled smoothly. Changes may be called amendments, modifications, add-ons, up-scopes, or down-scopes. Changes are a necessary aspect of

business for buyers, because of changes in their needs. Changes are opportunities either to increase or decrease revenues and/ or profitability for the seller.

- ***Invoices and payments:*** An efficient process must be developed for handling invoices and payments throughout contract performance. Few areas cause more concern to sellers than late payment. Buyers can realize savings by developing an efficient and timely payment process, because sellers are often willing to give discounts for early payment.

- ***Contract administration policies:*** Although the specific policies that will apply to contract administration depend on the parties, four policies are key: compliance with contract terms and conditions, effective internal and external communication and control, effective control of contract changes, and effective resolution of claims and disputes.

- ***Project Budget:*** The project budget is the amount of approved and authorized money needed to perform the work.

- ***Stakeholder Expectations:*** The stakeholder expectations are the written and unwritten requirements, goals, and desires of the key decision makers and end customers.

WHAT ARE THE RECOMMENDED TOOLS AND TECHNIQUES?

A variety of tools and techniques can be employed by the team in the performance management process for successfully managing project and contract performance. Which of these tools and techniques are employed depends, in part, on size and scale of the project, the knowledge and experience of the team members, and the organizational environment.

Performance Management Plan

A performance management plan should be prepared for virtually all project work. The PMP typically includes the following: project scope and objectives, integrated project team designation, roles and authority, responsibility assignment matrix, work breakdown structure, project schedule and critical milestones, and budget. Other aspects may include fielding and implementation plans, operational concepts, baseline establishment and tracking, problem escalation, and others, as warranted by the particular project.

Some of these aspects are worked out by applying the third discipline, Governance. Note that while all of these are essential elements of an earned value management system, the PMP establishes the foundation.

Case Study: CH2M Hill

CH2M Hill's strategic project management system emphasizes the consistent use of a standard project delivery process including team development, project chartering, work planning, endorsing, executing, managing change, and learning.

These overarching processes are totally integrated with the Project Management Institute's A Guide to the Project Management Body of Knowledge (PMBOK® Guide) and aligned to the ISO 10006 Global Standards for Project Management. These core processes are scalable and matched with Web-based technologies and tools that enable project managers to efficiently deliver products and provide accurate project status reporting to all levels of management – and clients.

The project manager development framework, along with this Web-based performance enhancement process, has been invaluable in the recognition and growth of project managers and their project delivery teams. Clear roles, responsibilities, and pathways for greater achievement have created a very attractive workplace environment.

CH2M Hill is considered a leader in environment restoration projects and credits much of their success to the ability to manage performance via their project management system.

Balanced Scorecard Method

The balanced scorecard (BSC) is a multidimensional framework for describing, implementing, and managing activities at all levels of an organization by linking objectives, initiatives, and measures to the organization's strategy. The scorecard provides an enterprise view of an organization's overall performance by integrating financial measures with other key performance indicators. The BSC is not a static list of measures, but a framework for implementing and aligning complex programs of change and for managing strategy-focused organizations.[1]

The BSC is a performance measurement methodology developed to provide organizations a simple framework for linking strategic goals to team and individual performance targets and requirements. The BSC method is currently used by more than 70 percent of the Fortune 1000 companies, according to a recent survey by the Bain Company. Plus, numerous U.S. government agencies have adopted the BSC method to help plan, link and communicate key aspects of performance. The BSC examines organizational performance in four critical areas: financial, customer, employee, and quality. See Figure 9-2 below.

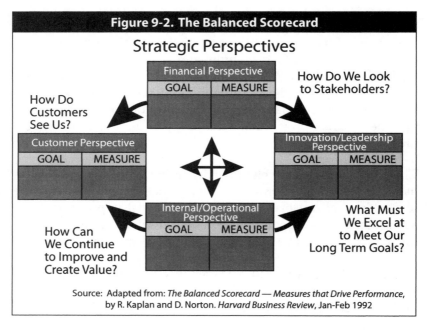

Figure 9-2. The Balanced Scorecard

Strategic Perspectives

Financial Perspective
GOAL / MEASURE

How Do We Look to Stakeholders?

How Do Customers See Us?

Customer Perspective
GOAL / MEASURE

Innovation/Leadership Perspective
GOAL / MEASURE

Internal/Operational Perspective
GOAL / MEASURE

What Must We Excel at to Meet Our Long Term Goals?

How Can We Continue to Improve and Create Value?

Source: Adapted from: *The Balanced Scorecard — Measures that Drive Performance,* by R. Kaplan and D. Norton. *Harvard Business Review,* Jan-Feb 1992

Pre-Performance (or Kick-Off) Conference

Before performance begins, the buyer and the seller should meet to discuss their joint management of the contract and project. The meeting should be formal; an agenda should be distributed in advance and minutes should be taken and distributed. At the meeting, the parties should review the contract terms and conditions and discuss who will do what. They also should establish protocols for written and oral communication and for performance measurement and reporting and discuss procedures for managing change and resolving differences. Buyer and seller key managers who will have performance responsibilities should attend the pre-performance conference. Important subcontractors also should be represented.

The pre-performance meeting "can help both agency and contractor personnel achieve a clear and mutual understanding of contract requirements and further establish the foundation for good communications and a win-win relationship."[2]

Payment System

Every contract must establish a clear invoicing and payment system or process. The buyer and seller must agree to whom invoices should be sent and what information is required. Sellers must submit proper invoices in a timely manner. Buyers should pay all invoices promptly. The payment system can be enhanced in its effectiveness if the contract includes performance incentives.

Quality Plan or Quality Assurance Surveillance Plan

The quality plan (usually developed by the seller) and/or quality assurance surveillance plan (usually developed, but certainly approved, by the buyer) serves as the baseline for the business outcomes/results expected to be achieved by the project. This document is the roadmap for measuring overall quality of the products and services. The quality plan must identify the measures, methods of measurement, and schedules the team will use to assess the quality of deliverable products or services. Under the Quality Plan, the project is evaluated on a regular basis to provide confidence that the project will satisfy the relevant quality standards. It should also be used to determine if specific project outcomes comply with required quality standards, and to identify ways to eliminate causes for unsatisfactory performance.

There are three major benchmarks against which performance should be measured: Are we delivering the project on time? Are we delivering the project within budget? Are we delivering results?

If the project is not producing the required outcomes, then the Government buyer and contractor teams must take corrective action to increase the effectiveness and efficiency of the project.

We recommend going a step further by posing the following question: Do the required results we are delivering help the agency achieve its overall goals and objectives? If the answer is "no" then the project objectives should be re-evaluated along with the performance measures and metrics to ensure that the strategic linkage is always kept in focus. One means to stay on focus is to adopt a matrix that aligns objectives, performance standards, acceptable quality level, measurement method, and incentives or disincentives, such as that shown in a Performance Matrix, Figure 9-3.

Keep in mind that good performance measures are:

■ Valid and objective – based upon reliable and accurate data, sources and methods,
■ Cost-effective – in terms of gathering and processing information,
■ Understandable – easy for decision-makers and stakeholders to use and act upon,
■ Tied to incentives – negative and positive.

Figure 9-3. Performance Matrix				
Business Objective	Performance Standard	Acceptable Quality Level (AQL)	Method of Surveillance	Incentives/ Disincentives for Exceeding or Not Meeting the AQL

Earned Value Management Systems (EVMS)

Earned value management systems (EVMS) – a term that replaces the 1960s' Cost/Schedule Control Systems Criteria (C/SCSC) – are widely accepted tools in industry for improving project management. A common operational definition of an EVMS is "the use of an integrated management system that coordinates work scope, schedule, and cost goals and objectively measures progress toward these goals."[3]

To standardize the use of EVMS across the government, the Department of Defense, General Services Administration, and National Aeronautics and Space Administration published a final rule amending the Federal Acquisition Regulation to implement EVMS policy in accordance with Office of Management and Budget (OMB) Circular A-11, part 7, and the supplement to part 7, the Capital Planning Guide. As a result, contracting officers, program managers, and contractors are now required to manage contracts by using earned value management systems for major acquisitions. Performance-based acquisition management requires the use of EVMS on those parts of the acquisition where developmental effort is required. The purpose of the EVMS rule is to give OMB the tool to require stricter budgetary discipline where it sees fit, even in lower dollar contracts.

The rule defers to agency policies regarding specific thresholds (dollar or otherwise) for when and on what acquisitions EVMS is to be implemented. The requirement to flow-down EVMS requirements from the prime contractor to its subcontractors will be governed by the same rules as those applied at the prime contractor level.[4]

The essential elements of an EVMS are: (1) organizing, (2) authorizing, (3) scheduling, (4) budgeting, and (5) performance measurement and analysis.

Organizing – Organizing the work is the initial task of an EVMS. In order to organize the work you must begin by creating a work breakdown structure (WBS), organizing an integrated project team, and developing a responsibility assignment matrix.

Authorizing – All work within a project should be described and authorized through a work authorization system. Work authorization ensures that performing organizations are specifically informed regarding their work scope, schedule for performance, budget, and charge number(s) for the work assigned to them. Work authorization is a formal process that can consist of various levels. Each level of authorization is agreed upon by the parties involved so that there is no question as to what is required.

Scheduling – The subjects of scheduling and budgeting are interrelated and iterative. In order to develop a time-phased budget plan, the schedule must be prepared first. Scheduling is the process of integrating activities and resources into a meaningful arrangement, depicting the timing of the critical activities that will satisfy the customer's requirements.

Budgeting – Budgeting is the process of distributing budgets "top down" to individual work segments.

Performance Measurement and Variance Analysis – Performance measurement for the functional managers, project control managers, and others consist of evaluating work packages' status calculated at the work package level. A comparison of the planned value (budgeted cost for work scheduled (BCWS)) to earned value (budgeted cost for work performed (BCWP)) is made to obtain the schedule variance, and a comparison of the BCWP to the actual costs for work performed (ACWP) is made to obtain the cost variance. Performance measurement provides a basis for management decisions by the project manager, management and, in some cases, the customer.

The results of EVMS monitoring produce cost and schedule performance data that are often displayed graphically to give the analyst and the manager a picture of the trends. An example, Figure 9-4, is provided below.

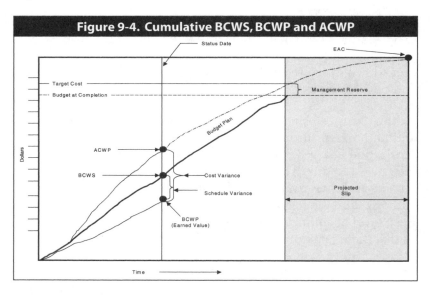

Figure 9-4. Cumulative BCWS, BCWP and ACWP

Change Control System

The only constant is change. Therefore, an effective process for managing change must be in place. "Managing change" means ensuring that changes are promptly identified, their effect is estimated and provided for, and are authorized—and that the other party is properly notified, compliance and impact are reported, compensation is provided, and the entire transaction is properly documented.

Best practices in contract and project change control include the following:

■ Ensure that only authorized people negotiate or agree to contract changes

■ Make an estimate of the effect of a change on cost and schedule, and gain approval for any additional expense and time before proceeding with any change

■ Notify project team members that they must promptly report (to the project manager of contract manager) any action or inaction by the other party to the contract that does not conform to the contract terms and conditions

■ Notify the other party in writing of any action or inaction by that party that is inconsistent with the established contract terms and conditions

■ Instruct team members to document and report in writing all actions taken to comply with authorized changes and the cost and time required to comply

■ Promptly seek compensation for increases in cost or time required to perform, and negotiate claims for such compensation from the other party in good faith

■ Document all changes in writing, and ensure that both parties have signed the contract; such written documentation should be completed before work under the change begins, if practical.

Contract Changes Clauses – Contracts, whether federal or commercial, frequently include a clause that authorizes the buyer to order the seller to conform with certain changes made at the buyer's discretion. Such clauses are called *change clauses.*

Documenting Change – Whenever the parties make a change in the contract, it is important to maintain the integrity of that document as trustworthy evidence of the terms and conditions of their agreement. Logically, a change will add terms and conditions, delete terms and conditions, or replace some terms and conditions with others. Thus, when modifying the contract, the parties should decide what words, numerals, symbols, or drawings must be added, deleted, or replaced in the contract document.

Parties to a contract will often discuss the change they want to make but fail to describe the change in the context of their contract document. After a few such changes, the document will no longer accurately describe the current status of their agreement, and the parties may dispute what the current terms really are. Such an occurrence should surprise no one, human communication and memory being what they are.

The best way to avoid this problem is to draft the language of the change carefully in the context of the contract document, ensuring that the new language describes the intent of the parties. This action should be taken before making any attempt to estimate the cost and schedule effect of any change or to perform the work. People sometimes argue that expediency demands that the work proceed before reaching agreement on the precise language of the change. However, this practice is likely to create confusion over just what changed and how. If the parties cannot reach agreement on the language of the change in a reasonable time, they probably are not in agreement about the nature of the change and should not proceed.

Modification of the Contract Document – One party will have the original copy of the contract. The other party will usually have a duplicate original. These originals should remain with the contract manager, or in the contracts department or legal office.

When parties agree to change the contract, they should never alter the original documents. Instead, they should prepare modification documents that describe the contract changes. These changes can generally be described in two ways: First, the modification document can include substitute pages in which deleted original language is stricken out and new or replacement language is inserted in italics. Second, minor changes can be described in "pen and ink" instructions that strike out certain words and add others.

Copies of each modification should be distributed promptly to all project team members who have a copy of the original document. The project manager, contract manager, and other key team members should maintain a personal conformed working copy of the contract. This copy should be kept in a loose-leaf binder or electronic database so that pages can be replaced easily. The conformed working copy should be altered as necessary to reflect the current status of the agreement between the parties.

Changes should be incorporated promptly. Each team member should always keep the conformed working copy readily available and bring it to meetings. The contract manager should periodically check to ensure that each team member's conformed working copy is up-to-date.

Effect of the Change on Price and Schedule – After the parties are in precise agreement as to how the contract was modified, they should try to estimate the cost and schedule impact of the change. They can do this independently, but the most effective approach is to develop the estimate together, as a team, working out the details and their differences in the process. If the parties are open and honest with one another, this approach can save time and give them greater insight into the real effect of the change on cost and schedule. A well-developed work breakdown structure and project schedule graphic can be of enormous value to this process. Work may proceed based either on an estimate of the cost and schedule impact, with a limit on the parties' obligations, or on a firm-fixed adjustment.

Authorization of Performance Under the Change – After the parties have agreed to the change and to either an estimate of the impact on cost and schedule or a final price adjustment, the buyer should provide the seller with written authorization to proceed with the work as changed. The easiest way to accomplish this objective is to prepare, sign, and distribute a modification document. If this approach will take too long, a letter or other form of written documentation will suffice. The authorization should include a description of the change, its effective date, and a description of any limits on the obligations of the parties.

Case Study: Bechtel Corporation

Bechtel, a world-class construction firm, has achieved an outstanding industry reputation for excellent contract fulfillment. Bechtel is well known for clearly defining roles and responsibilities for its company, its supply chain partners, and its customers. Bechtel ensures each contract has a well-defined Statement of Work, a project delivery schedule, a contract change management process, clear acceptance criteria, and an effective invoice and payment process. Bechtel has demonstrated for many years the importance of tying contract administration and project management together to leverage higher performance and higher profitability.

At Bechtel, managing change means ensuring that changes are authorized, their effect is estimated and provided for, they are promptly identified, the other party is properly notified, compliance and impact are reported, compensation is provided, and the entire transaction is properly documented.

Dispute management system

No one should be surprised when, from time to time, contracting parties find themselves in disagreement about the correct interpretation of contract terms and conditions. Most such disagreements are minor and are resolved without too much difficulty. Occasionally, however, the parties will find themselves entangled in a seemingly intractable controversy. Try as they might, they cannot resolve their differences. If the dispute goes unresolved for too long, one or both of the parties may threaten, or even initiate, litigation.

Litigation is time consuming, costly, and risky. No one can ever be entirely sure of its result. It rarely results in a truly satisfactory resolution of a dispute, and it sours commercial relationships. For these reasons, it should be avoided. One goal of business managers and contract managers should be to resolve disputes without litigation whenever possible.

The keys to effective dispute resolution are as follows:
- Recognize that contract documents are not perfect
- Keep larger objectives in mind
- Focus on the facts
- Depersonalize the issues
- Be willing to make reasonable compromises.

When disputes become intractable, seeking the opinion of an impartial third party can sometimes help. When this approach is formal, and the third party's decision is binding on the parties, it is called *arbitration.* Some companies include a clause in their contracts that makes arbitration the mandatory means of resolving disputes. Such a clause might read as follows:

Disputes

Should any dispute occur between the parties arising from or related to this Agreement, or their rights and responsibilities to each other, the matter shall be settled and determined by arbitration under the then current rules of the American Arbitration Association.

> The arbitration shall be conducted by a single arbitrator, the decision and award of the arbitrator shall be final and binding, and the award so rendered may be entered in any court having jurisdiction thereof. The language to be used in the arbitral proceeding shall be English.

In an international contract, the "Disputes" clause may be modified to provide for an international forum. The clause might read—

> The arbitral tribunal shall be composed of three (3) arbitrators who shall be appointed by the Chairman of the Royal Arbitration Institute of the Stockholm Chamber of Commerce. The arbitration process will be more formal than ordinary negotiation between the parties (which may be represented by attorneys), but it will be less formal than court proceedings.

Analysis of Performance

Observed and collected information about project performance must be analyzed to determine whether the project is proceeding as planned. The analyst compares actual performance to performance goals to determine whether any variances exist. An analyst who discovers a variance between actual and expected performance must determine several things: Is it significant? What was its cause? Was it a one-time failure, or is it a continuing problem? What type of corrective action would be most effective?

Variance analysis must be timely, particularly when the information is obtained through reports. That information is already old by the time it is received. Delays in analyzing its significance may allow poor performance to deteriorate further, perhaps beyond hope of effective corrective action. Acting promptly is particularly important during the early phases of contract performance, when corrective action is likely to have the greatest effect.

It is not uncommon for project managers and contract managers to collect reams of information that sit in their in-baskets and file cabinets, never put to use. When a project has gone badly, a review of information in the project files frequently shows that there were warning signs—reports, meeting minutes, letters, memos—but that they were unnoticed or ignored. Often, several people, perhaps a variety of business managers, share responsibility for monitoring performance. In these instances, the project manager and contract manager must take steps to ensure that those people promptly analyze the information, report their findings, and take corrective action.

Corrective Action – When the project manager and contract manager discover a significant variance between actual and expected performance, they must take corrective action if possible. They must identify the cause of the problem and determine a solution that will not only eliminate it as a source of future difficulty, but also correct the effect it has already had, if possible. If the effect cannot be corrected, the parties may need to negotiate a change to the contract, with compensation to the injured party, if appropriate.

Follow-Up – After corrective action has been taken or is under way, the project manager and contract manager must determine whether it has had or is having the desired effect. If not, further action may be needed. Throughout this corrective action and follow-up process, the parties must keep each other informed about what is going on. Effective communication between the parties is essential to avoid misunderstandings and disputes when things are not going according to plan. The party taking corrective action must make every effort to let the other party know that it is aware of the problem and is addressing it seriously. Sometimes this step is more important than the corrective action itself.

What are the Desired Outputs?

At the completion of a project the desired outputs typically include the following: on-time delivery; on-budget performance; quality products, services, and/or solutions that met or exceeded all requirements; and high customer satisfaction. As a summary, the following are 30 best practices to improve performance results.

Best Practices for Buyers and Sellers: 30 Actions to Improve Performance Results

- Select a project manager and contract manager to ensure that your organization does what it proposed to do
- Form an Integrated Project Team
- Read and analyze the contract
- Develop a project performance plan

- Comply with contract terms and conditions
- Maintain effective communication and control
- Control contract changes
- Resolve claims and disputes promptly and dispassionately
- Use negotiation or arbitration, not litigation, to resolve disputes
- Develop a work breakdown structure to assist in planning and assigning work
- Conduct pre-performance conferences
- Measure, monitor, and track performance using an Earned Value Management System (EVMS)
- Manage the invoice and payment process
- Report on progress internally and externally
- Identify variances between planned versus actual performance
- Be sure to follow up on all corrective actions
- Appoint authorized people to negotiate contract changes and document the authorized representatives in the contract
- Enforce contract terms and conditions
- Provide copies of the contract to all affected organizations
- Maintain conformed copies of the contract
- Understand the effects of change on cost, schedule, and quality
- Document all communication—use telephone and correspondence logs
- Prepare internal and external meeting minutes
- Apply EVM cost and schedule variance analysis
- Ensure completion of work
- Document lessons learned and share them throughout your organization
- Communicate, communicate, communicate!
- Clarify team member roles and responsibilities
- Provide leadership support to the team throughout the performance management process
- Ensure that leadership understands the performance management process and how it can improve business relationships from beginning to end.

SUMMARY

So in retrospect, in this chapter, we discussed the Sixth Discipline of Performance-Based Project Management (PBPM), Performance Management. Specifically, we reviewed the key inputs, recommended tools and techniques, and desired outputs of the performance management process. We discussed in some detail numerous valuable tools and techniques to improve performance management, including: performance management plans, contract analysis, quality plan, payment system, change control system, earned value management system (EVMS), and dispute management system. All of the aforementioned tools and techniques are designed and intended to help organizations meet or exceed their mission results, when properly structured, tailored to the project, and intelligently implemented.

QUESTIONS TO CONSIDER

1. Does your organization develop performance management plans?

2. How effectively does your organization develop and implement a Quality Plan or Quality Assurance Surveillance Plan?

3. Has your organization effectively implemented an Earned Value Management System (EVMS)?

4. Does your organization have an effective Contract Change Control System?

5. How well does your organization manage disputes?

Endnotes

[1]   Acquisition Solutions *Advisory*,"The Balanced Scorecard in Acquisition," by Bob Welch, April 2000.

[2]   Seven Steps to Performance Based Acquisition, Step 7, at *http://www.acquisition. gov/comp/seven_steps/step7_add.html*.

[3]   *http://www.acq.osd.mil/pm/faqs/faq.htm*.

[4]   Acquisition Solutions Virtual Acquisition Office™ Daily Acquisition News Article, "FAC2005-11 implements EVMS policy and provides reference to emergency acquisitions," July 5, 2006.

THE FUTURE ACQUISITION WORKFORCE

By: Shirl Nelson

INTRODUCTION

In the world of federal government buying, the quality and quantity of the people, or as it is called, the "acquisition workforce," has a direct bearing on whether and how well an agency achieves its mission. This acquisition workforce includes all those people who have responsibility for defining requirements, measuring contract performance, providing technical and management direction, and contracting.[1] Clearly, these are mission-critical functions. However, in the federal government, decades of downsizing, under-investing, under-training, and under-hiring are colliding with workforce demographics and the "now portable" federal retirement system to create a crisis in the numbers, experience, domain expertise, and management and leadership skills across the federal acquisition workforce.

Using the disciplined Performance-Based Project Management (PBPM) approach described in the preceding chapters is an excellent action plan for getting results, but every action plan relies on people to achieve the results. Consider the impact of these statistics on the acquisition workforce:

- U.S. Federal agencies spend over $400 billion a year for a wide range of products and services to meet their mission needs.[2]
- Government purchasing increased by nearly 75 percent between 2000 and 2005.[3]
- Service contracting accounts for more than 60 percent of total obligations each year[4] and services are increasingly complex in nature, demanding higher skill levels to execute.
- The size of the federal acquisition workforce has declined by nearly 50 percent since the 1990s.[5]
- The Acquisition Workforce includes more than a dozen specialized career fields, including: Contract Specialist, Program Manager, Cost/Price Analyst, Logistics Manager, Government Property Manager, Contracting Officer's Technical Representative (COTR), Engineers, Computer Scientists, Industrial Specialists, and others.

As federal agencies address the challenges of increasing complexity and workload in the face of a diminished acquisition workforce, they have an opportunity to re-engineer their business processes on an enterprise scale and rebuild and shape that workforce into a value-added, results-oriented entity that is fully embraced by the agency in addressing its strategic challenges. The challenges are formidable: recognize the need for and then influence cultural transformation, make strategic linkages, establish governance structures and processes, communicate effectively, manage risk, and manage performance to achieve results. Whether leading a team responsible for a specific acquisition, managing a portfolio of acquisitions, or leading an organization, acquisition personnel need greater skills than has been the case in the past.

What, then, should the acquisition workforce of the future look like, in order to meet these challenges? This chapter will examine this question from both an individual and organizational perspective. Let's start with the individual.

WHAT WILL THE ACQUISITION PROFESSIONAL OF THE FUTURE LOOK LIKE?

The acquisition professional who is succeeding now and will succeed in the future will have a combination of hard and soft skills that in sum add up to the potential to become an agency leader, not just of acquisition, but of any agency function or even the agency overall. If you are contemplating starting or advancing your career in federal acquisition, this is what you need to know: Acquisition encompasses both the contracting and requirements sides of agency programs. The roles for both contracting and requirements (program) officials engaged in acquisition are changing. The role of the contracting professional is shifting from that of a conventional-process, risk-adverse contracting officer to one of a results-oriented strategic business advisor viewed as a key member of agency leadership and program teams. The role of the program official is shifting from that of a task director to that of one who articulates mission objectives and manages contractors' performance for results.

WHAT IS A "STRATEGIC BUSINESS ADVISOR?"

Strategic business advisor is our term of art—or vision—for the role acquisition professionals ideally should play in the planning, execution, and management of federal business deals (whether contracts or grants). It is an evolution of a concept of a "business broker" or "business manager" that has long been discussed as the optimal role for a contracting professional. But in our view, it is *more*. It is a professional who is more than a broker, more than a manager: It is someone who exhibits all three characteristics of the three-word term. Strategic business advisors are—

Business Broker

I believe a Business Broker is an individual who provides business acumen to the acquisition team. This person will possess technical knowledge, expertise in applying sound business principles, and good judgment that creates innovative sourcing solutions to achieve agency or organizational needs. The focus is on creating solutions.

— Deidre Lee, then-Director of Defense Procurement,
in the July-August 2000 issue of Program Management

Business Manager

Compounding the uncertainty of the future environment is the changing role of the acquisition professional from merely a purchaser or process manager to a business manager. Uncertainty is also caused by an increased focus on performance and outcomes, which requires greater integration of functions such as acquisition, financial management, and program management. In order to make this transition, acquisition workers will need to acquire an entirely new set of skills and knowledge, according to the agency officials with whom we spoke. For example, in addition to having a firm understanding of contracting rules and processes, acquisition workers will need to be adept at consulting and communicating with line managers, and they will need to be able to analyze business problems, identify different alternatives in purchasing goods or services, and assist in developing strategies in the early stages of the acquisition.

Finally, a deeper understanding of market conditions, industry trends, and the technical details of the commodities and services being procured will be required.

— GAO, Acquisition Workforce: Status of Agency Efforts
to Address Future Needs (report # 03-55), December 2002.

Strategic. They are centric to meeting agency objectives and are knowledgeable of the marketplace, the current environment, the customer's needs, the objectives, the parameters, the constraints— and the possibilities. They consider all the variables and all stakeholders and are strategic in crafting a business (acquisition) plan for execution and management.

TEN

Mission and Business-minded. They concentrate on exploring business arrangements that focus all parties on the same objectives— as partners working toward a shared goal—and apply available financial resources and compensation arrangements in a manner that incentives "win/win" and generates high performance.

Advisors. They are customer-focused, applying their knowledge, experience, skills, and business focus to achieving the customer's goals, while providing the highest levels of performance, clarity, focus, and trust.

In other words, it is more than just an elegant label; it is a way of thinking about the acquisition role that is results-focused, not contract focused. There are pressures that work against this. Given workload and tradition, a contracting officer or contract specialist can fall into the trap of being too narrowly focused on putting together a solicitation to respond to a deadline, then awarding a contract and moving on to the next one. This can lead contracting officers and specialists to disengagement from the mission of the agency and the results of the contract.

The business advisor, on the other hand, seeks and is able to think and behave strategically and connect his or her activities to agency goals and objectives. The business advisor is aware of the budgetary, political, regulatory, and technical environment of the program and knows the market space. The business advisor knows a lot about companies that can fulfill program needs and the economics and business rules of the marketplace in which those companies operate. The business advisor knows what drives bid/no-bid decisions in the marketplace and, therefore, how to generate effective competition.

Like their counterparts in the private sector, federal business advisors either are—or are poised to be—an integral part of their agency's senior management. More than 60 percent of industry executives say that by 2015, their companies will have a chief procurement officer who will report to the chief executive officer and set a strategic course for purchasing company-wide.[6]

Is the role of program officials changing?

Yes, this role is also shifting. Traditional program officials in the federal government have often seen their role narrowly and far removed from "acquisition," even when the programs managed are largely fulfilled through contract support. Contracting has been something that the procurement office does and the traditional program manager's role was to tag the contracting officer with a requisition who is then "it." But program managers, long considered part of the acquisition workforce in the Department of Defense, are now recognized as such throughout government.[7] Perhaps more importantly, more of these key players themselves recognize their role and responsibilities in the acquisition arena.

As a counterpart to the Department of Defense (DoD) Defense Acquisition Workforce Improvement Act (DAWIA) of 1991 which provided Contracting and Program Management certification requirements, the Federal Acquisition Institute (FAI) issued Federal Acquisition Certification requirements for Program and Project Managers (FAC-P/PM) in January 2007.[8] FAC-P/PM establishes certification, training, and continuous learning requirements for entry, journeyman, and expert level program and project managers engaged in acquisition. The intent of these certification requirements is to provide a results-oriented, competency-based program to support achievement of an agency's mission through sound acquisition program and project management. Further, acquisition officials—contracting, program, and other players—are now routinely forming integrated project teams (IPTs) in recognition of the interdependence of their roles in achieving results.

WHAT DOES IT TAKE TO BE A STRATEGIC BUSINESS ADVISOR?

In addition to specific competencies related to the acquisition of commodity or service areas involved, we believe the following characteristics are essential for the strategic business advisor role:

- Strategic thinking
- Business acumen
- Results focus
- Comfort with ambiguity
- Learning agility
- Interpersonal skills
- Self-confidence and a sense of humor
- Persuasiveness; an ability to find common ground
- Teamwork
- A quest for continuous improvement.

These characteristics are a compilation of our thoughts and those of thought leaders and research organizations over the past several years. In the early nineties, the National Association of Purchasing Management's (NAPM), now called the Institute of Supply Management (ISM), Center for Advanced Purchasing Studies (CAPS) conducted a survey of 700 chief purchasing officers of large U.S. firms, concluding that purchasing professionals should have the following key skills: interpersonal communication, customer focus, decision-making ability, analytical and negotiation skills, conflict resolution skills, flexibility, problem-solving skills, ability to influence and persuade, and computer literacy.

To function as key figures in their agencies, the acquisition leaders of tomorrow must be able to think and behave strategically and connect their activities to agency goals and objectives. That ability is supported by acumen in areas such as:

- **Business and organizational savvy:** In addition to the standard contracting official's professional development, the acquisition professional of the future must be accomplished at articulating and conveying personal and organizational values and skilled in organizational politics, networking, and follow-through as aids to "getting things done" effectively, efficiently, and with business acumen. While these "soft" skills often are overlooked in traditional acquisition training programs, they have a significant impact for success.

- **Market expertise:** As acquisition professionals become more like partners in achieving the agency's strategic objectives, they will benefit by developing greater market expertise to establish benchmarks for performance, build supplier relationships, and incorporate new technological developments more easily that solve the agency's problems.

- **Understanding economics and market forces:** From the perspective of the acquisition professional, economics is not just an "ivory tower" concept, but an element affecting contract negotiations and supplier relationships. Understanding market forces allows the business advisor to optimize market conditions and understand the supplier's need to meet a certain return on investment. This awareness provides a lens through which to view possibilities for constructing a win-win partnership.

- **Commodity or service area focus:** As industry has done for years, federal acquisition professionals increasingly must learn about and specialize in their customers' products and service needs: What is the reason for the acquisition? What mission must it fulfill? What companies operate the commodities involved? What is the future of the underlying technologies? What is the commercial marketplace for similar commodities? Not only will their in-depth commodity focus help them gain an enhanced understanding of the customers' needs, it also will help ensure a supplier base for future acquisitions and quality industry partnerships.

- **Comfort with technology and continuous learning:** Acquisition professionals will find that technology not only is a central feature in meeting the specific needs of their agencies' missions, but also is a tool to enable the sharing of information, as well as more efficient work processes. But technology is only an enabler. Acquisition professionals need to embrace the capture and reuse of knowledge to shorten learning curves and improve performance. Further, they must be visionary on an enterprise level. Work processes must be streamlined from an operational perspective, reducing labor-intensive activities and resulting in more efficient flow of information, decisions, and controls and enhancing overall results. Understanding and planning for technology is increasingly important in implementing commercial systems across functional processes, where acquisition professionals must play vitally needed leadership roles.

As contracting professionals transition to a business advisor role for their respective agencies, they should broaden their standard professional development with a more strategic focus. They no longer are just fulfilling orders from their customers. They need to operate with an enterprise view. Finally, their perspective should be that of a more strategic partner with the agency's other senior executives, to anticipate and prepare for changes in the mission, marketplace, and technological world.

WHAT SHOULD ORGANIZATION LEADERS DO TO BEGIN TRANSFORMING THEIR CONTRACTING STAFF TO STRATEGIC BUSINESS ADVISORS?

First, be aware that the concern for the acquisition workforce is shared at the highest levels of government. Congress, the Government Accountability Office (GAO) and the Office of Federal Procurement Policy (OFPP) recognize that agency leaders need help and support in rebuilding their acquisition workforce. As this book goes to press, proposed legislation in both houses of Congress is focused on the acquisition workforce[9] The GAO has identified Human Capital as a cornerstone for building successful organizations and identified critical success factors for that cornerstone, reflected in Table 10-1 below.

Table 10-1. GAO Framework Elements and Success Factors for Human Capital		
Cornerstone	**Element**	**Critical Success Factor**
Human Capital	Valuing and Investing in the Acquisition Workforce	Commitment to Human Capital Management Role of the Human Capital Function
	Strategic Human Capital Management	Integration and Alignment Data-Driven Human Capital Decisions
	Acquiring, Developing, and Retaining Talent	Targeted Investments in People Human Capital Approaches Tailored to Meet Organizational Needs
	Creating Results-Oriented Organizational Cultures	Empowerment and Inclusiveness Unit and Individual Performance Linked to Organizational Goals
Source: Acquisition Solutions™ Advisory, "Rising to the Challenge with Strategic Business Advisors," January 2007		

The Office of Federal Procurement Policy has undertaken a survey to identify key competency gaps and develop targeted training and strategies to meet broad acquisition workforce needs. We recommend that agencies implement their own, more focused assessments, and put strategic human capital plans in place to begin the transformation.

WHAT SHOULD AGENCY ASSESSMENTS FOCUS ON?

Ideally, agencies should focus first on the strategic alignment of the acquisition function with agency goals, then on the processes and policies of their acquisition organizations to meet those goals, finally on the skills and size of their acquisition workforce needed to execute those goals. However, the magnitude and severity of the current workforce crisis have pushed many leaders to first address the size of their workforce, with strategic alignment, quality, processes, and policies–hopefully–to follow later. All too often "later" never comes.

Therefore, we recommend to agency managers an overarching assessment strategy with seven organizational dimensions to be analyzed and transformed in separate, but related, initiatives:

- Strategic Validation and Management
- Workload Assessment and Management
- Workforce Assessment and Management
- Client and Stakeholder Identification and Relationship Management
- Supplier Identification and Relationship Management
- Financial Management Analysis and Improvements
- Policy and Process Review and Improvement.

The key to the ultimate success of these initiatives will require the introduction of knowledge management techniques, appropriate technology insertion, and new tools and techniques for management–including the use of performance metrics.

The work should be tackled in three phases: assessment and establishment of baseline, development of strategy and plan, and implementation. The time required depends on the organization, available resources, and degree of commitment and support.

Figure 10-1

PHASE I	PHASE II	PHASE III
STRATEGY VALIDATION		
• Align to Vision and Strategy • Create Awareness of the High-Level Strategy and Plan	• Define Overall Strategy • Risk and Mitigation Strategies	• Implementation
WORKLOAD ASSESSMENT & MANAGEMENT		
• Execute Spend Analysis • Perform Contract Management Review	• ID Integrated Procurement Tools and Techniques	• Implementation
WORKFORCE ASSESSMENT & MANAGEMENT		
• Perform Staffing and Workload Analysis	• Develop Development Plan • Establish Position Description • Institute Recruiting Strategy	• Implementation
CLIENT & STAKEHOLDER IDENTIFICATION & RELATIONSHIP MANAGEMENT		
• Conduct Client and Stakeholder Profiles	• Develop Communication Plan • Create Governance Model	• Implementation
SUPPLIER ID & RELATIONSHIP MANAGEMENT		
• Profile Supplier Base • Perform Supplier Interviews	• Conduct Market Research Analysis • Develop Communication Plan	• Implementation
FINANCIAL MANAGEMENT ANALYSIS & IMPROVEMENT		
• ID High-Level Spent Savings • ID Bulk Buy Opportunities	• Develop Spend Reduction Strategy • Capture Metrics	• Implementation
POLICY & PROCESS REVIEW & IMPROVEMENT		
• Perform Policy Gap Analysis • Conduct Policy Management	• Introduce New Policy & Procedures • Deploy Knowledge Convergence Framework	• Implementation
PROGRAM MANAGEMENT		
METRICS		
STRATEGIC SOURCING		

2006 © Acquisition Solutions, Inc.

Ultimately, you want recommendations for improving the acquisition function which not only address regulatory compliance issues, but *equally important* lead to increased customer satisfaction and a workforce enabled to provide mission-critical acquisition support in the most efficient manner. Continually think ahead and ask critical questions, such as:

- How will changes in policies and processes enhance the stated objectives of the organization? This story must be sound and convincing.
- What resources must be committed to develop tools required by new initiatives? Employees' hands should not be tied to old tools and expected to be efficient at new processes.
- What training and employee development is essential for new processes and strategies to take hold? Committing resources to professional development will demonstrate management commitment to new processes.
- Are performance appraisal systems, rewards and recognition consistent with the new behaviors desired? What gets measured gets done; these must be in alignment.

■ What communication strategies are necessary to ensure that rhetoric is consistent with leadership behavior? Employees must believe management is walking the talk.

As the assessment is undertaken, the organization's leaders need to validate their vision and strategy, then evaluate the current organization, identify strengths and weaknesses by comparing current practices to the best practices in the field. Next we suggest developing a strategy to improve the operation, with proposed key performance indicators for managing and measuring the acquisition function. We suggest four foundational steps for accomplishing these objectives.

1. ***Baseline Assessment:*** Perform a baseline assessment by collecting and analyzing appropriate and relevant data, including workforce skills; analyzing contract files; conducting interviews; and reviewing relevant acquisition and program documentation. This will provide an understanding of the existing conditions and issues at the organization.

2. ***Best Practices Evaluation:*** Based on a thorough understanding of the maturity of the current acquisition function, collect and identify relevant industry and sector best practices, then determine their potential applicability to the environment.

3. ***Gap Analysis:*** Perform a gap analysis by comparing the current state of the organization's processes to selected best practices. Based on this comparison, identify, as necessary, skills gaps of the acquisition personnel, key issues and concerns, critical areas needing improvement, risks, and management decisions necessary to ensure organizational success.

4. ***Develop Report and Implementation Plan:*** Based on a comprehensive understanding of the environment, existing processes and practices, relevant best practices, and associated gaps and risks, develop the report with findings, conclusions, and recommendations. Be sure the conclusions and recommendations give consideration to increasing productivity, address customer satisfaction, and support the agency's mission. Once management has adopted the recommendations, develop an implementation plan for use in guiding improvement efforts. The implementation plan should include: (a) high level objec-

tives to be pursued in each of the best practices areas where there are recommendations; (b) prioritized actions to accomplish the objective; (c) identification of implementation leads (agency offices/positions); and (d) scheduled timeframes for implementation.

The agency's objectives must be kept in mind throughout. The following is an example of a typical map for alignment of agency objectives with our recommended process:

Table 10-2	
Desired Agency Outcomes	**Study Process**
Obtain a comprehensive understanding of the acquisition mission and the challenges and issues facing the organization. This may include reviewing the current organizational alignment, reviewing acquisition policies and processes, and measuring and managing performance. Examine existing policies, procedures, and general guidance and recommend updates/revisions and new polices needed.	Step 1: Baseline Assessment
Obtain information on external best practices in acquisition and identify ways that the acquisition organization can utilize these procedures in their own operation. Benchmark the organization's acquisition practices against other government agencies and commercial companies.	Step 2: Best Practices Evaluation
Evaluate existing practices for meeting performance based contracting and small business goals and identify ways to improve performance in these areas. Analyze and identify specific areas of risk. Evaluate the composition of staff resources and seek suggestions for identifying the effective use of these resources. Conduct a skills gap analysis of acquisition personnel.	Step 3: Gap Analysis
Obtain recommendations for process improvement opportunities. Plan for support services as needed to implement of recommended improvements.	Step 4: Develop Report and Implementa- tion Plan
Source: Acquisition Solutions, Inc.	

WHAT IS THE "RIGHT" SIZE FOR AN AGENCY'S ACQUISITION WORKFORCE?

The answer is, "It depends." There are some widely accepted benchmarks for determining the numbers, such as cost-to-spend, spend-per-employee, and spend-to-budget ratios. Let us explain.

The *cost-to-spend ratio* is an indication of the efficiency of an organization's operating costs; that is, the lower the ratio, the less an agency spends for getting its procurement budget obligated.

The *spend-per-employee ratio* is an indication of the productivity of an organization's employees; that is, the higher the value, the more procurement budget is obligated by each employee.

Finally, the *spend-to-budget ratio* is considered a measure of the impact that procurement operations have on an agency's mission; that is, the higher the percentage, the greater the impact. But these ratios should not be viewed in a vacuum. There is more to the story.

WHAT IS "THE REST OF THE STORY?"

Each organization is unique in mission, culture, design, and process maturity. Even though other agencies' and industries' ratios and statistics provide comparison benchmarks, we have not found sufficient correlation between agencies' obligations and transactions to rely on benchmarks exclusively for estimating staffing resources. We therefore recommend agencies consider their places on the maturity curve in various management areas before deciding on their staffing requirements. The quality of the contracting, professionalism of the staff, management effectiveness, organization placement, and checks and balances all have a material effect on the quantity of staff required and grade structure to be considered. Consider the following areas and questions when contemplating resources.

Organizational structure and authority: What level of procurement authority exists within the contracting operation? To what extent are contracting officer warrants provided to promote delegation to the optimum level? Are warrants issued only to those qualified with proper education, training, certifications, and experience? Is the organization operating on parity with its customers?

Span of control: What is the supervisory span of control (supervisor to nonsupervisor ratio)? Is it representative of a control-oriented environment, where production and compliance are valued, or of a more empowered environment preferred by higher educated, self-motivated leaders of change? Where managers function more as enablers of change rather than reviewers of work, the span is larger. When day-to-day supervision is necessary, a much lower employee-to-supervisor ratio is in order.

Policies and procedures: Are proper and current policies, procedures, and processes in place to ensure efficient use of resources? Is it necessary to continuously issue policy letters that fall outside the normal policy infrastructure? Are employees confused about organizational practices in certain areas, necessitating peer-to-peer collaboration on how others are approaching a practice?

Quality and internal management controls: Are acquisition plans, source selection plans, and award decisions reviewed at the appropriate levels? Is an acquisition system review process in place to detect process issues, and does management act on it to improve quality? Have negative audit findings detected quality issues and lack of internal controls?

Level of automation and reliability of information: Do managers have the right information at the right time to make decisions? Are procurement processes automated and integrated with other agency systems to improve efficiency?

Training and development programs: Is the staff well trained to perform routine tasks (such as simplified purchases) that clearly require less supervision? Is the staff trained in emerging practices that provide value-added services to the customer? Is the training budget sufficient to ensure continuous training of the staff?

Knowledge management practices: Do managers encourage the capture, sharing, and re-use of knowledge? Have they integrated into their daily practice the learning before, during, and after techniques that we describe in our chapter on Knowledge Leadership? Have they hired or trained staff to serve as "knowledge engineers" to facilitate effective teams for project execution?

Trend analysis: Are budgets increasing or decreasing? Is the spend per employee rising or falling? Is the cost-to-spend increasing or decreasing? Are the number of transactions increasing per employee or decreasing? Organizations should track data over time and analyze trends to synthesize the information in the context of these other characteristics of an organization's maturity.

Complexity and acquisition practices: Are acquisitions becoming more complex? Are major systems being replaced, requiring more sophisticated thinking? Is the agency truly embracing key initiatives, such as performance-based acquisition and strategic sourcing?

Customer alignment and teaming: Does the agency embrace integrated project teaming, where the contacting officer functions as the "deputy for acquisition," or is a "throw it over the wall'" mentality frustrating customers and diminishing organizational efficiency?

Employee morale and turnover: Are employees leaving at an unusually high rate? Is stress affecting morale? The point is clear. If the current staff is poorly trained, or the organization has yet to streamline its procurement processes and embrace new reforms, or a micromanagement environment exists, or morale is low, hiring at the same level of proficiency is a poor use of resources. Making the case for more full-time equivalents (FTEs) should be done in an environment where the mission and management of the agency are enhanced by excellence. Throwing more bodies into an otherwise unhealthy situation is ineffective at best. In other words, determining the "right size" of the workforce should be done as part of a comprehensive organizational assessment that includes strategic acquisition human capital planning.

Strategic acquisition human capital planning ... what is it and how do I do it?

Strategic acquisition human capital planning is a systematic approach to assessing (and reassessing) long-term acquisition workforce resources needed to fulfill an agency mission. It requires that agencies align their organizations, project needs, assess current resources, identify the gaps, then develop and implement a plan for filling those gaps.[10] It should be a goal to develop this business advisor expertise through all intern, exchange, training, human capital planning, or other initiatives to improve the acquisition workforce.

What guiding principles for recruitment and retention would we recommend?

We recommend five core principles that should guide improvements in acquisition workforce recruitment and retention:

- My organization will foster the development of business advisor competencies.
- My organization will offer a clear and inviting path for quality college graduates to enter and rise to *leadership positions.*

- My organization will base requirements for entry on academic and performance merit.
- My organization will offer ample opportunity for participation in structured training, communities of practice, and enriching assignments broader than the instant job.
- My organization will value the sharing of knowledge and experience across functional areas and with the private sector to bring insight and business acumen to the federal acquisition practice.

CONCLUSION

The roles of acquisition professionals are shifting. Regardless of one's particular role in the cycle, skilled professionals understand the planning, budgeting, requirements, contracting, and performance management aspects of acquisition. They understand the marketplace and the economic drivers that generate effective competition of ideas and price. In short, they bring "business advisor" skills to the acquisition table. Their managers are leaders of organizations who encourage learning and knowledge sharing. They embrace best practices and continually seek to improve the organization with policies and procedures that facilitate effective operations. Together, the team—the managers and members of the "big A" Acquisition workforce recognize the interdependencies of today's multi-sector environment, and view contractors as partners whose success performing a contract is the agency's success.

The world in which we live is changing at an ever-increasing rate, creating a demand for both velocity and flawless execution. Business advisors need discipline—the six disciplines—to perform in this environment. So here is the challenge: transform culturally, link strategically, govern, communicate, manage risk, and measure performance ... to get results!

Questions to Consider

1. What actions is your organization taking to improve knowledge transfer?

2. Does your organization have an effective intern program for all critical acquisition roles?

3. Does your organization create and effectively implement succession planning at all levels for key acquisition positions?

Endnotes

[1] "Acquisition" is defined by the Services Acquisition Reform Act of 2003 (P.L. 108-136) to include defining requirements, measuring contract performance, providing technical and management direction, and contracting.

[2] FPDS-NG statistics for 2006.

[3] Acquisition Advisory Panel final panel working draft report, December 2006.

[4] Ibid.

[5] Ibid.

[6] "The New Face of Purchasing," *The Economist*, April 2005.

[7] Office of Federal Procurement Policy Letter 05-01, Developing and Managing the Acquisition Workforce, April 15, 2005.

[8] Under authority of 41 U.S.C. 401, et seq and OFPP Policy Letter 05-01.

[9] Examples: Accountability in Contracting Act (H.R. 1362), Accountability in Government Contracting Act of 2007 (S.680), and Department of Homeland Security Procurement Improvement Act of 2007 (H.R. 803).

[10] OPM maintains statistics on the federal workforce at *http://www.fedscape.gov*.

NATIONAL ACADEMY OF PUBLIC ADMINISTRATION MULTISECTOR WORKFORCE: MISSION- CRITICAL AREAS

Source: http://www.napawash.org/about_academy/MultisectorWork-force12-13-05.pdf

ACCOUNTABILITY

■ Who should be held accountable for the accomplishment of federal missions performed by workers from other sectors?

■ How can they be held accountable?

■ How do we assign roles and responsibilities to the federal manager to ensure accountability for the performance of the multisector workforce, not just that of federal employees?

■ Traditional tools for accountability include the budget, contracts, and grants. Are there other tools, resources/influencers, case studies, information sharing and/or best practices to improve government performance and our leadership in this area?

■ Are there different accountability issues when the work is being done by other levels of government?

■ How do we align the policies affecting the multisector workforce through the roles of human resources specialists, acquisition specialists and managers?

■ Where contractors have been engaged with inadequate analysis of the task or specification of performance requirements, rewards and sanctions, what consequences and follow-up occurred?

■ What systems can be developed to address contract administration accountability issues such as identifying contractors who have been temporarily suspended and documenting issues of waste, fraud, and abuse?

■ How do we evaluate outsourcing activities to validate cost savings over time?

■ How do we improve post award accountability?

■ Are peer reviews or other models applicable to government accountability?

ACQUISITION

- How do we develop, implement and evaluate contracting vehicles to ensure agencies have needed competencies, obtain surge capacity, acquire needed flexibility and resolve specific issues?

- How do we streamline and improve the use of regulations as a mechanism for accountability?

- What is the impact of the government no longer directly employing the workforce that is conducting new research and development and building new innovations and technology?

- Has the definition of "inherently governmental" shifted – however subtly – so that contractor personnel are performing functions that ought to be performed by federal employees?

- Are there emerging best practices for acquisition by government agencies that use multisector workforces?

HUMAN CAPITAL AND MANAGEMENT

- What special skills are needed to manage a multisector workforce?

- What tools, systems and best practices exist to model effective management of the third party workforce?

- How do we address the need to improve the skills and competencies of the current employee and supervisory workforce so they are able to work with, as well as oversee and manage, the multisector workforce?

- Can federal performance management systems – including pay-for–performance systems – be designed to support coordination, oversight, and management of the multisector workforce?

- How do we develop and institutionalize a workforce planning system that accommodates a view of the entire multisector workforce needed to accomplish the mission, not just the federal component?

- How can government assure that third party workforces will provide the skills and competencies needed?

- How do we build project management capacity and acquisition skills needed to improve our management of federal contracts?

- How do we sustain core competencies to ensure effective project management, oversight and termination of contracts if necessary?

- What role do labor unions play as representatives of employees in the workforce, when half of that workforce is a contractor workforce and the balance are federal employees?

- How can we help managers avoid pitfalls such as supervising contractors and allowing contractors to provide personal services?

- What is the impact of contractors supervising other contractors on behalf of the federal government or supervising federal employees?

SOCIAL EQUITY AND VALUES

- What is the impact of contracting out on the actual delivery of services, i.e., who is served, who is not served, and how is the public need being met?

- What impact does contracting out have on the values and goals inherent in the federal government's treatment of its own workforce? If there is a negative impact, can or should it be remedied?

- How do we address the fact that the federal government has traditionally addressed equity values in employment, whereas many who are increasingly doing the work of government do not?

- Does it make a difference that federal employees take the Oath of Office, while contractor employees do not?

LEGAL AND GOVERNANCE ISSUES

■ Is there a changing paradigm of government management from a governance focus to an entrepreneurial (business) focus? How does the growth of the multisector workforce contribute to that phenomenon?

■ What is the impact on our Constitutional system and administrative law norms when government activities are performed by a multisector workforce through various contracting and grant vehicles?

■ How do we stay true to the public purpose of administrative laws such as freedom of information, open meetings, enforcement proceedings, avoidance of conflict of interest and public participation with respect to the activities of government contractors? How do we address this issue across Federal, state and local government lines?

■ What constitutes a "coherent framework of laws," management principles, and organizational practices to assure that government officials have the tools they need to account for the work of the government?

■ Contract specialists currently consider costs, contracting vehicles, and performance, but do not always consider the impact of constitutional and administrative law as part of the equation for determining the business arrangement. Should we work to ensure that the contracting specialists have knowledge of and consider the impact of constitutional and administrative law norms when making contracting decisions?

■ How can we determine the best alternatives for establishing a governance structure for the multisector workforce?

ORGANIZATIONAL CULTURE

■ Considering meetings, teambuilding techniques, labor management partnerships, training, communication vehicles, work group design, performance and feedback instruments, which are the most effective techniques for organizing and integrating the multisector workforce around an agency's culture and mission?

- Currently, government managers have limited knowledge of the rules and norms by which the private sector operates. What do managers need to know about the private sector to be effective managers of the multisector workforce? What do managers need to teach the private sector about government?

- How do we develop systems that allow for comparable recognition of effort by all parts of the multisector workforce? Currently, agencies may directly award traditional cash and honorary recognition for accomplishing results only to federal employees.

- The research indicates that differences in work status (contractor vs. federal employee) have similar weight in the workplace as other demographics of race, gender, etc. What is the impact of this finding?

- What happens to the culture of an organization when different employees are working under different pay and benefit plans?

- How has the relationship between the employee and the federal government changed as a result of increased use of the multisector workforce? How has it changed the "psychological contract" with respect to employee concerns in areas such as job security, recruitment, retention and pay for performance?

- What measures and performance management systems should be employed with the multisector workforce to ensure effectiveness of that workforce?

APPENDIX  **B**

TEMPLATES/
EXAMPLES

BYLAWS FOR PROGRAM EXECUTIVE BOARD

ARTICLE 1 — NAME, PURPOSE

Section 1:
Name of Organization: Program Executive Board

Section 2:
Mission & Vision - Purpose

ARTICLE II — BOARD

Section 1:
Composition & Size

Examples:
- Members are appointed by President and will number no fewer than three nor more than seven.
- Membership is defined by the roles of the key stakeholders: e.g., Chief Information Officer, Chief Acquisition Officer, Legal Officer, Program Director,
- Membership may include Senior Vice President of Company "X" (Contractor), who might serve either as full member or as an advisory member.

Section 2:
Board Role

Example:
The Board is responsible for overall direction of the Program and delegates responsibility for day-to-day operations to the Program Manager and Committees

Section 3:
Meetings.

Example:
- The Board shall meet at least quarterly, at an agreed upon time and place
- The Board will convene at Key Program Milestones, to consider whether or not the program should proceed to the next phase.

Section 4:
Terms.

Examples
- All Board members shall serve three-year terms, but are eligible for re-appointment. However, no board member shall serve more than two three-year terms. The first Board will include members with one and two-year terms to begin staggered terms.
- Board members will serve for the planned duration of the project.

Section 5:
Quorum.

Example:
- A quorum must be attended by at least forty percent of the Board members before business can be transacted or motions made or passed.

Section 6:
Notice.

Example
- An official Board meeting requires that each Board member have written/email notice two weeks in advance.

Section 7:
Officers and Duties.

Examples:
- There shall be three officers of the Board consisting of a Chair, a first Vice-Chair, and Executive Secretary.
- The **Chair** shall convene regularly scheduled Board meetings, shall preside or arrange for other members of the executive committee to preside at each meeting in the following order: first Vice-Chair, Executive Secretary .
- The **Vice-Chair** will chair committees on special subjects as designated by the board.
- The Program Manager, in the role of **Executive Secretary** shall be responsible for keeping records of Board actions, in- cluding overseeing the taking of minutes at all board meetings, sending out meeting announcements, distributing copies of

minutes and the agenda to each Board member, and assuring that corporate records are maintained.

Section 8:
Resignation, Termination and Absences.

Examples:
- Resignation from the Board must be in writing and received by the Secretary.
- If a stakeholder organization notifies the Board that their representative who serves on the Board no longer represents the stakeholder organization, the person is no longer eligible.
- A Board member shall be dropped for excess absences from the Board if he or she has three unexcused absences from Board meetings in a year.
- A Board member may not delegate membership to a subordinate; however a subordinate may attend as a non voting member in the absence of the board member.

Section 9:
Special Meetings.

ARTICLE III - COMMITTEES

Section 1:
The Board may create committees as needed.

Example:
- There shall be xxx standing committees -
- The Chair appoints all committee chairs.
- Committee chairs must be members of the Board.

ARTICLE VI – PROGRAM MANAGER AND STAFF

Section 1:
Program Manager.

Example:
- The Program Manager has day-to-day responsibility for the project, including carrying out the organizations goals and Board policy/decisions..
- The Program Manager will serve as Executive Secretary of the Board

- Will attend all Board meetings,
- Will report on the progress of the Program,
- Will answer questions of Board members and
- Will carry out the duties described in the job description. The Board can designate other duties as necessary.

ARTICLE VII - AMENDMENTS

Section 1:
These Bylaws may be amended when necessary by a two-thirds majority of the Board. Proposed amendments must be submitted to the Secretary to be sent out with regular Board announcements.

APPENDIX

EARNED VALUE MANAGEMENT SYSTEM (EVMS)

Earned Value Management Systems (EVMS) are widely accepted tools in industry for improving project management. A common operational definition of an Earned Value Management (EVMS) is "the use of an integrated management system that coordinates work scope, schedule, and cost goals and objectively measures progress toward these goals." The term EVMS replaces the old term used since the 1960's Cost/Schedule Control Systems Criteria (C/SCSC).

To standardize the use of EVMS across the government, the Department of Defense, General Services Administration, and National Aeronautics and Space Administration published a final rule amending the Federal Acquisition Regulation to implement earned value management system (EVMS) policy in accordance with Office of Management and Budget (OMB) Circular A-11, part 7, and the supplement to part 7, the Capital Planning Guide. Effective July 5, 2006, contracting officers, program managers, and contractors are required to manage contracts by using earned value management systems for major acquisitions. Performance-based acquisition management requires the use of EVMS on those parts of the acquisition where developmental effort is required. The purpose of the EVMS rule is to give OMB the tool to require stricter budgetary discipline where it sees fit, even in a lower dollar contract.

The rule defers to agency policies regarding specific thresholds (dollar or otherwise) for when and on what acquisitions EVMS is to be implemented. The requirement to flow-down EVMS requirements from the prime contractor to its subcontractors will be governed by the same rules as those applied at the prime contractor level.[1]

The essential elements of an EVMS are the following: (1) Organizing, (2) Authorizing, (3) Scheduling, (4) Budgeting, and (5) Performance Measurement and Analysis.

Organizing – Organizing the work is the initial task of an EVMS. In order to organize the work you must begin by creating a Work Breakdown Structure, organizing an Integrated Project Team, and developing a Responsibility Assignment Matrix.

Work Breakdown Structure (WBS) – the WBS provides the framework for the organization of the contract effort. It is an indentured listing of all of the products (e.g. hardware, software, services, and data) to be furnished by the seller. It is used as the basis for all contract planning, scheduling, and budgeting; cost accumulation; and performance reporting throughout the entire period of project performance.

Integrated Project Team (IPT) – The Integrated Project Team structure reflects the organization required to support the project. The project manager is responsible to ensure the cost, schedule, and technical management of the project. The Project Manager draws upon the functional groups to accomplish the work through the assignment of responsibility to appropriate managers.

Responsibility Assignment Matrix (RAM) – The RAM ties the work that is required by the WBS elements to the organization responsible for accomplishing the assigned tasks. The intersection of the WBS with the integrated project team structure.

Authorizing – All work within a project should be described and authorized through a work authorization system. Work authorization ensures that performing organizations are specifically informed regarding their work scope, schedule for performance, budget, and charge number(s) for the work assigned to them. Work authorization is a formal process that can consist of various levels. Each level of authorization is agreed upon by the parties involved so that there is no question as to what is required.

The document involved in work authorization should be maintained in a current status throughout the lifecycle of the contract as revisions take place.

Customer Authorization – Customer authorization is comprised of:
- Basic contract
- Contract change notices
- Engineering change notices

Figure 1 shows a typical work authorization process.

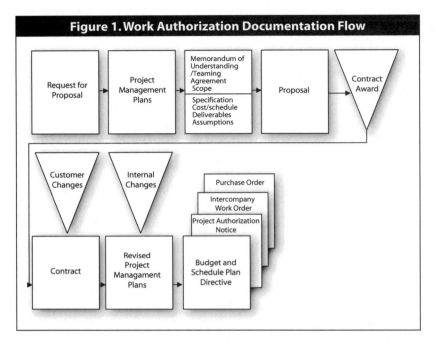

Figure 1. Work Authorization Documentation Flow

Scheduling – The subjects of scheduling and budgeting are inter-related and iterative. In order to develop a time-phased budget plan, the schedule must be prepared first. Scheduling is the process of integrating activities and resources into a meaningful arrangement, depicting the timing of the critical activities that will satisfy the customer's requirements.

Project Scheduling – Project scheduling is a logical time-phasing of the activities that are necessary to accomplish the entire project scope. It is the most important tool for cost and schedule:

- Planning,
- Tracking,
- Analysis of variances, and
- Reporting of project performance.

Each activity in the network is characterized by scope, logical relationships, duration, and resources.

Figure 2 shows the steps required to build a project schedule.

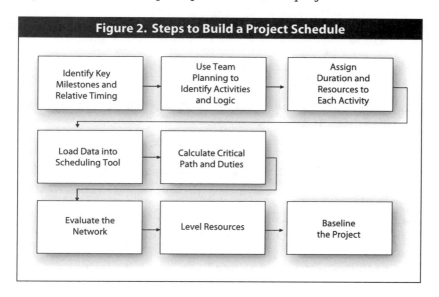

Figure 2. Steps to Build a Project Schedule

Scheduling Definitions

■ Milestone – An event of particular significance that has no duration.

■ Activity – Something that occurs over time; work that must be accomplished, also referred to as a "task."

■ Sequential – Activities that are performed in sequence or right after each other.

■ Concurrent/Parallel – Two or more activities that are performed at the same time or that overlap.

Scheduling Terms

Finish-to-Start: The predecessor activity must be completed before the successor activity can begin.

Start-to-Start: The predecessor activity must begin before the successor activity can begin.

Finish-to-Finish: The predecessor activity must end before the successor activity can end.

Lag – Any schedule time delay between two tasks. Lags can be positive or negative.

Critical Path – Longest continuous sequence of tasks through the network, given the underlying relationships that will affect the project end date.

Float – Difference between the time available (when tasks can start/finish) and the time necessary (when tasks must start/finish).

Schedule Outputs – There are three basic scheduling outputs:
- Network diagram,
- Gantt chart,
- Resource histogram.

Once the project schedule is complete, the cost/schedule performance baseline is established. Figure 3 shows the steps required to establish the baseline and track and analyze performance on the project.

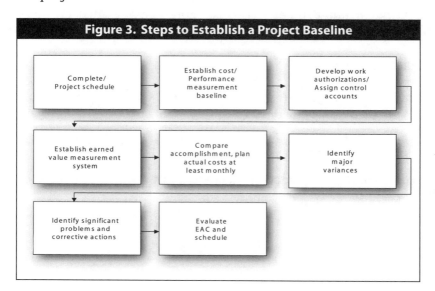

Figure 3. Steps to Establish a Project Baseline

Figure 4 is a diagram of a Performance Measurement System:

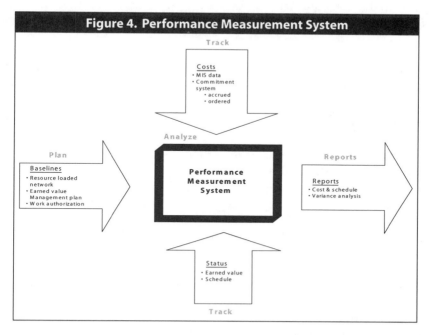

Figure 4. Performance Measurement System

Budgeting – Budgeting is the process of distributing budgets to individual work segments. The following top-down illustration, Figure 5, gives the overview of the relationships.

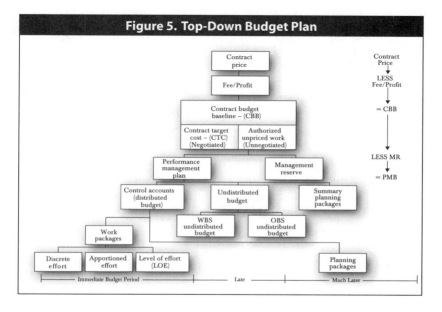

Figure 5. Top-Down Budget Plan

Total Allocated Budget (TAB) – The TAB is the sum of all budgets allocated to the contract. It is the same as the Contract Budget Base (CBB) unless an Over Target Baseline (OTB) has been established. (See *Changes* for an explanation of OTB).

Contract Budget Base – The CBB is the sum of the Contract Target Cost (CTC) plus the estimated cost of any authorized unpriced (not yet negotiated) work. It is made up of the Performance Measurement Baseline and management reserve.

Performance Measurement Baseline (PMB) – The PMB is the time-phased budget plan against which contract performance is measured. It is composed of the budgets assigned to Control Accounts and Undistributed Budget. It equals the TAB minus management reserve.

Management Reserve – This is a budget amount that is set aside to provide for unforeseen, within-contract scope requirements.

Control Account – The control account is the focal point for planning, monitoring, and controlling project work as it represents the work within a single WBS element, and is the responsibility of a single organizational unit.

Virtually all aspects of the Performance Management System come together at the control account level, including budgets, schedules, work assignments, cost collection, progress assessment, problem identification, corrective actions, and Estimate at Completion (EAC) development. Day-to-day management is performed at the control account level.

The level selected for the establishment of a control account must be carefully considered to ensure that work is properly defined into manageable units with responsibilities clearly delineated.

Undistributed Budget – This is the budget that is applicable to a specific contract effort, but that has not yet been distributed to the WBS elements. Undistributed budget is intended to serve only as a temporary holding account until the budget is properly distributed.

Summary Planning Packages – Summary planning packages are used to plan time-phased budgets for far-term work that cannot practically be planned in full detail.

Work Packages – A work package is a detailed job that is established by the functional manager for accomplishing work within a control account.

A work package has these characteristics:

- Represents units of work (activities or tasks) at the levels where the work is performed.
- Is clearly distinct from all other work packages, and is the responsibility of a single organizational element.
- Has scheduled start and completion dates (with interim milestones, if applicable) that are representatives of physical task accomplishment.
- Has a budget or assigned value expressed in terms of dollars, labor hours, or other measurable units.
- Has a duration that is relatively short, unless it is subdivided by discrete milestones to permit objective measurement of work performed.
- Has a schedule that is integrated with all other activities occurring on the project.
- Has a unique Earned Value technique, either discrete, apportioned effort, or Level of Effort (LOE).

Performance Measurement & Variance Analysis - Performance measurement for the functional managers, project control managers, and others consist of evaluating work packages status calculated at the work package level. A comparison of the planned value (Budgeted Cost for Work Scheduled (BCWS)) to Earned Value (Budgeted Cost for Work Performed (BCWP)) is made to obtain the schedule variance, and a comparison of the BCWP to the Actual Costs (ACWP) is made to obtain the cost variance. Performance measurement provides a basis for management decisions by the project manager, management and, in some cases, the customer.

Performance Measurement

Performance measurement provides:
- Work progress status,
- Relationship of planned cost and actual cost to actual accomplishment,
- Valid, timely, auditable data,
- Basis for the EAC.

Elements required to measure project progress and status are:
- Work package schedule status,
- BCWS or the planned expenditure,
- BCWP or Earned Value,
- ACWP or MIS costs and accruals.

Control account/work packages:
- Measurable work and related event status form the basis for determining progress for BCWP calculations.
- BCWP measurements at summary WBS levels result from accumulating BCWP upward through the control account from the work package levels.
- Within each control account, the inclusion of LOE is kept to a minimum to prevent distortion of the total BCWP.

Calculation methods used for measuring work package performance are:
- Short work packages (2 months or less) may use the measured effort or formula method, e.g., 0-100%, where a status can be applied each month.
- Longer work packages (over 2 months) should have milestones assigned. The milestones are then statused monthly for the lifecycle of the work package.
- In manufacturing, work packages may use the earned standards or equivalent units method to measure performance based on the manufacturing work measurement system output.
- Effort that can be measured in direct proportion to other discrete work may be measured as apportioned effort work packages. Apportioned effort is used primarily in manufacturing.
- Sustained efforts are planned using the LOE Earned Value method. The Earned Value for LOE work packages is equal to the time-phased plan (BCWS).
- The measurement method used depends on an analysis of the work to be performed in the work package. Whichever method is selected for planning (BCWS) must also be used for determining progress (BCWP).

Estimated to Complete (ETC) Preparation.

To develop an ETC, the CAM must consider and analyze:
- Cumulative ACWP/ordered commitments,
- Schedule status,

- BCWP to date,
- Remaining control account scope of work,
- Previous ETC,
- Historical data,
- Required resources by type,
- Projected cost and schedule efficiency,
- Future actions,
- Approved contract changes.

The functional managers or Control Account Managers (CAM) prepare the ETC as required by the Project Manager.

EAC Preparation. The ETC is then summarized to all necessary reporting levels, added to the ACWP and commitments, and reported to Corporate management and the customer, as appropriate.

A bottoms-up EAC should be prepared quarterly for all contracts.

Estimate at Completion – The EAC is the estimated cost at the end of the project. See Figure 6. It is composed of the cost of what has been accomplished and the estimated cost of the remaining work. The following graph, Figure 7, illustrates the two primary components of the EAC.

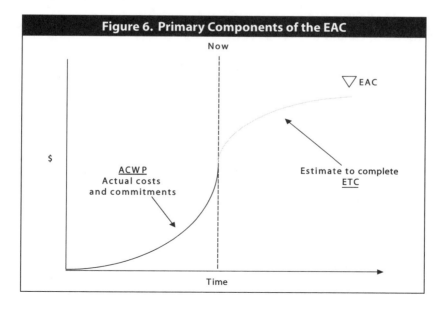

Figure 6. Primary Components of the EAC

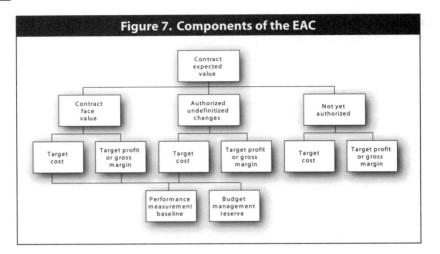

Figure 7. Components of the EAC

Revenues at Completion – Revenues at completion are the total revenues anticipated on the contract at the completion of the project. Revenues at completion are composed of the EAC and the EAC profit.

EAC Profit – EAC profit is the profit expected to be achieved at the completion of the project.

Estimate at Completion (EAC) – EAC is the cost of work performed to date plus the estimated cost of all remaining work on the project. The EAC is made up of four components: ACWP, open commitments, ETC, and hard reserves.

Actual Cost of Work Performed (ACWP) – ACWP is the cost of work performed to date plus accruals.

Estimate to Complete (ETC) – ETC is the estimated cost of the remaining work on the project.

If performance measurement produces schedule or cost variances in excess of pre-established thresholds, the cause must be determined. The functional managers or CAM is responsible for the analysis of the control account and understanding trends that indicate potential future problems.

Variance Calculations – There are three types of variances: schedule variances, cost variances, and Variances of Completion (VACs). They are calculated as follows:

$$SV = BCWP - BCWS$$

$$CV = BCWP - ACWP$$

$$VAC = BAC - EAC$$

Variance Thresholds -Variance analysis is required when one or more of the variances exceeds the threshold established for the project. Variance thresholds are defined by a percentage or a dollar amount, or a combination of the two. The latter method is usually more appropriate since it eliminates very small variances from the analysis requirement. The thresholds are generally established by the Sector, but may be provided by the customer.

Variance Analysis Operation

The Variance Analysis Reports (VARs) provide current period, cumulative, and at-completion data. CAMs provide VARs for control accounts that have a schedule variance, cost variance, or VAC that exceeds the established thresholds.

The CAM completes the VAR by providing a description of the cause of the variance, its impact on the control account and other elements of the project, the corrective action to be taken, and any follow-up on previous actions taken.

The VAR is submitted through the appropriate project channels for approval.

The Project Manager uses the control account VARs to report project status to upper management.

The Project Manager has a continuing responsibility for monitoring corrective actions.

Periodic, formal project reviews, scheduling meetings, and staff meetings serve as forums for variance trend analysis and corrective action monitoring.

Legend – This legend is applicable to the formulae and charts that follow:

BCWS	Budgeted Cost for Work Scheduled
BCWP	Budgeted Cost for Work Performed
ACWP	Actual Cost of Work Performed
BAC	Budget at Completion
ETC	Estimate to Complete
EAC	Estimate at Completion

Cost Performance Analysis Basic Formulae

Cost Variance

$$CV = BCWP - ACWP$$

Cost Variance %

$$CV\% = \frac{CV}{BCWP} \times 100$$

Cost Performance Index

$$CPI = \frac{BCWP}{ACWP}$$

To Complete Performance Index

$$TCPI = BAC - BCWPcum$$

$$EAC - ACWPcum$$

Schedule Variance

$$SV = BCWP - BCWS$$

Schedule Variance %

$$SV\% = \frac{SV}{BCWP} \times 100$$

Schedule Performance Index

$$SP I = \frac{BCWP}{BCWS}$$

Schedule Variance in Months

$$SV\ months = \frac{SV\ cum}{BCWP\ current\ period}$$

Percent Spent

$$\%\ spent = \frac{ACWP\ cum}{BAC^*} \times 100$$

Percent Complete

$$\%\ complete = \frac{BCWP\ cum}{BAC^*}$$

*EAC, PMB, CBB, or TAB may also be used.

Statistical Examples

Independent EAC

The basic formulae are:

$$EAC1 = ACWPcum + (BAC - BCWP\ cum)$$

$$EAC2 = \frac{BAC}{CPIe}$$

$$EAC3 = [(BAC - BCWP)/(CPI \times SPI)] + ACWP$$

Variance at Completion %u

$$VAC\% = \frac{VAC \times 100}{BAC}$$

Budget/Earned Rate

$$\text{E/B Rate} = \frac{\text{BCWP dollars}}{\text{BCWP hours}}$$

Actual Rate

$$\text{Actual Rate} = \frac{\text{ACWP dollars}}{\text{ACWP hours}}$$

Rate Variance

$$\text{Rate Variance} = (\text{B/E Rate} - \text{Actual Rate}) \times \text{Actual Hours}$$

To-Go-Rate

$$\text{To-Go Rate} = \frac{\text{ETC dollars}}{\text{ETC hours}}$$

Efficiency Variance

$$\text{Efficiency Variance} = (\text{BCWP hours} - \text{ACWP hours}) \times \text{B/E Rate}$$

Price Variance

$$\text{PV} = (\text{Planned/Earned Price} - \text{Actual Price}) \times \text{Actual Quantity}$$

Usage Variance

$$\text{UV} = (\text{Planned/Earned Quantity} - \text{Actual Quantity}) \times \text{Earned Price}$$

EVMS Graphical Trend Analysis – Cost and schedule performance data are often displayed graphically to give the analyst and the manager a picture of the trends. The two most common displays are shown here. These graphs can be used for a control account, an organization, a WBS element, or the entire project.

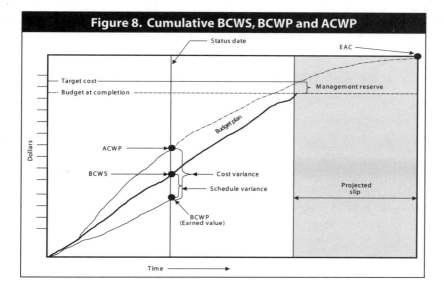

Figure 8. Cumulative BCWS, BCWP and ACWP

1   Acquisition Solutions Virtual Acquisition Office™ Daily Acquisition News Article, "FAC2005-11 implements EVMS policy and provides reference to emergency acquisitions," July 5, 2006.

BIBLIOGRAPHY

Atkinson, William, Beyond the Basics, PM Network Magazine, May 2003, Project Management Institute.

Badgerow, Dana B., Gregory A. Garrett, Dominic F. DiClementi, and Barbara M. Weaver, Managing Contracts for Peak Performance, Vienna, Va.: National Contract Management Association, 1990.

Bonaldo, Guy, Interview with Business 2.0 Magazine, Business Intelligence, February 2003.

Barkley, Bruce T., and Saylor, James H., Customer Driven Project Management: A New Paradigm in Total Quality Implementation, New York: McGraw-Hill, 1993.

Bossidy, Larry and Charan, Ram, Confronting Realty: Doing What Matters to Get Things Right, New York: Crown Business, 2004.

Bruce, David L., Norby, Marlys, and Ramos, Victor, Guide to the Contract Management Body of Knowledge (CMBOK), 1st ed., Vienna, VA: National Contract Management Association, 2002.

Cleland, David I., Project Management: Strategic Design and Implementation, New York: McGraw-Hill, 1994.

Cleland, David I., and King, William R., Project Management Handbook, 2nd ed., New York: Van Nostrand Reinhold, 1988.

Collins, Jim, Good to Great: Why Some Companies Make the Leap...and Others Don't, New York: Harper Collins, 2001.

Coulson-Thomas, Colin, Creating the Global Company, New York: McGraw-Hill, 1992.

Covey, Stephen R., The Seven Habits of Highly Effective People, New York: Simon and Schuster, Inc., 1989.

Fisher, Roger, Elizabeth Kopelman, and Andrea K. Schneider, Beyond Machiavelli: Tools for Coping with Conflict, Cambridge: Harvard University Press, 1994.

Freed, Richard C., Romano, Joe, and Freed, Shervin, Writing Winning Business Proposals, New York: McGraw - Hill, 2003.

Garrett, Gregory A., Achieving Customer Loyalty, Contract Management Magazine, August 2002, National Contract Management Association.

Garrett, Gregory A., Performance-Based Acquisition: Pathways to Excellence, McLean, VA,NCMA 2005.

Garrett, Gregory A., World-Class Contracting: How Winning Companies Build Successful Partnerships in the e-Business Age, 4th ed., Chicago, CCH, 2006.

Garrett, Gregory A., Managing Complex Outsourced Projects, Chicago, CCH, 2004.

Garrett, Gregory A., Contract Negotiations: Skills, Tools, & Best Practices, Chicago, CCH, 2005.

Garrett, Gregory A. and Bunnik, Ed, Creating a World-Class PM Organization, PM Network Magazine, September 2000, Project Management Institute.

Garrett, Gregory A., and Kipke, Reginald J., The Capture Management Life-Cycle: Winning More Business, Chicago, CCH, 2003.

Garrett, Gregory A. and Rendon, Rene G., Contract Management Organizational Assessment Tools, McLean, VA, NCMA, 2005.

Gates, Bill, Business @ The Speed of Thought: Using a Digital Nervous System, New York: Warner Books USA, 1999.

Harris, Phillip R., and Robert T. Moran, Managing Cultural Differences, Houston, Gulf Publishing Company, 1996.

Hassan H. and Blackwell R., Global Marketing, New York: Harcourt Brace Publishing, 1994.

Horton, Sharon, Creating and Using Supplier Scorecards, Contract Management Magazine, McLean, VA, NCMA, September 2004, pgs. 22-25.

Kantin, Bob, Sales Proposals Kit for Dummies, New York: Hungry Minds, 2001.

Kerzner, Harold, In Search of Excellence in Project Management, New York, Van Nostrand Reinhold, 1998.

Kirk, Dorthy, Managing Expectations, PM Network Magazine, August 2000, Project Management Institute.

Lewis, James P., Mastering Project Management: Applying Advanced Concepts of Systems Thinking, Control and Evaluation, Resource Allocation, New York, McGraw-Hill, 1998.

Liker, Jeffrey K. and Choi, Thomas Y., Building Deep Supplier Relationships, Harvard Business Review, Boston, MA December 2004, pgs. 104-113.

McFarlane, Eileen Luhta, Developing International Proposals in a Virtual Environment, Journal of the Association of Proposal Management, Spring 2000, Association of Proposal Management Professionals.

Monroe, Kent B., Pricing: Making Profitable Decisions, 2nd ed., New York: McGraw-Hill Publishing Company, 1990.

Moran, J. and Riesenberger M., The Global Challenge, New York: McGraw-Hill, 1994.

The National Contract Management Association, The Desktop Guide to Basic Contracting Terms, 4th ed., Vienna, Virginia, 1994.

O'Connell, Brian, B2B.com: Cashing-in on the Business-to-Business E-commerce Bonanza, Holbrook, Massachusetts: Adams Media Corp., 2000.

Ohmae, Kenichi, The Borderless World: Power and Strategy in the Interlinked Economy, New York: Harper Collins Pubs., Inc., 1991.

Ohmae, Kenichi, The Evolving Global Economy, Boston, MA: Harvard Business School Press, 1995.

Patterson, Shirley, Supply Base Optimization and Integrated Supply Chain Management, Contract Management Magazine, McLean, VA, NCMA, January 2005, pgs. 24-35.

Peterson, Marissa, Sun Microsystems: Leveraging Project Management Expertise, PM Network Magazine, January 2003, Project Management Institute.

Project Management Institute Standards Committee, A Guide to the Project Management Body of Knowledge, Upper Darby, Pa., Project Management Institute, 2001.

Reichheld, Frederick F., The Loyalty Effect, Boston, Harvard Business School Press, 1996.

Tichy, Noel, The Leadership Engine, New York, Harper Business Press, 1997.

Webster' Dictionary, The New Lexicon of the English Language, New York, Lexicon Publications, Inc., 1989.

Wilson, Greg, Proposal Automation Tools, Journal of the Association of Proposal Management, Spring/Summer 2002, Association of Proposal Management Professionals.

GLOSSARY

acceptance
(1) The taking and receiving of anything in good part, and as if it were a tacit agreement to a preceding act, which might have been defeated or avoided if such acceptance had not been made. (2) Agreement to the terms offered in a contract. An acceptance must be communicated, and (in common law) it must be the mirror image of the offer.

acquisition cost
The money invested up front to bring in new customers.

acquisition plan
A plan that serves as the basis for initiating the individual contracting actions necessary to acquire a system or support a program.

acquisition strategy
The conceptual framework for conducting systems acquisition. It encompasses the broad concepts and objectives that direct and control the overall development, production, and deployment of a system.

act of God
An inevitable, accidental, or extraordinary event that cannot be foreseen and guarded against, such as lightning, tornadoes, or earthquakes.

actual authority

> The power that the principal intentionally confers on the agent or allows the agent to believe he or she possesses.

actual damages

> See *compensatory damages*.

affidavit

> A written and signed statement sworn to under oath.

agency

> A relationship that exists when there is a delegation of authority to perform all acts connected within a particular trade, business, or company. It gives authority to the agent to act in all matters relating to the business of the principal.

agent

> An employee (usually a contract manager) empowered to bind his or her organization legally in contract negotiations.

allowable cost

> A cost that is reasonable, allocable, and within accepted standards, or otherwise conforms to generally accepted accounting principles, specific limitations or exclusions, or agreed-on terms between contractual parties.

alternative dispute resolution

> Any procedure that is used, in lieu of litigation, to resolve issues in controversy, including, but not limited to, settlement negotiations, conciliation, facilitation, mediation, fact-finding, mini-trials, and arbitration.

amortization

> Process of spreading the cost of an intangible asset over the expected useful life of the asset.

apparent authority

> The power that the principal permits the perceived agent to exercise, although not actually granted.

as is

A contract phrase referring to the condition of property to be sold or leased; generally pertains to a disclaimer of liability; property sold in as-is condition is generally not guaranteed.

assign

To convey or transfer to another, as to assign property, rights, or interests to another.

assignment

The transfer of property by an assignor to an assignee.

audits

The systematic examination of records and documents and/ or the securing of other evidence by confirmation, physical inspection, or otherwise, for one or more of the following purposes: determining the propriety or legality of proposed or completed transactions; ascertaining whether all transactions have been recorded and are reflected accurately in accounts; determining the existence of recorded assets and inclusiveness of recorded liabilities; determining the accuracy of financial or statistical statements or reports and the fairness of the facts they represent; determining the degree of compliance with established policies and procedures in terms of financial transactions and business management; and appraising an account system and making recommendations concerning it.

base profit

The money a company is paid by a customer, which exceeds the company's cost.

best value

The best trade-off between competing factors for a particular purchase requirement. The key to successful best-value contracting is consideration of life-cycle costs, including the use of quantitative as well as qualitative techniques to measure price and technical performance trade-offs between various proposals. The best-value concept applies to acquisitions in which price or price-related factors are *not* the primary determinant of who receives the contract award.

bid

An offer in response to an invitation for bids (IFB).

bid development

All of the work activities required to design and price the product and service solution and accurately articulate this in a proposal for a customer.

bid phase

The period of time a seller of goods and/or services uses to develop a bid/proposal, conduct internal bid reviews, and obtain stakeholder approval to submit a bid/proposal.

bilateral contract

A contract formed if an offer states that acceptance requires only for the accepting party to promise to perform. In contrast, a *unilateral contract* is formed if an offer requires actual performance for acceptance.

bond

A written instrument executed by a seller and a second party (the surety or sureties) to ensure fulfillment of the principal's obligations to a third party (the obligee or buyer) identified in the bond. If the principal's obligations are not met, the bond ensures payment, to the extent stipulated, of any loss sustained by the obligee.

breach of contract

(1) The failure, without legal excuse, to perform any promise that forms the whole or part of a contract. (2) The ending of a contract that occurs when one or both of the parties fail to keep their promises; this could lead to arbitration or litigation.

buyer

The party contracting for goods and/or services with one or more sellers.

cancellation

The withdrawal of the requirement to purchase goods and/or services by the buyer.

capture management

The art and science of winning more business.

capture management life cycle

The art and science of winning more business throughout the entire business cycle.

capture project plan

A document or game plan of who needs to do what, when, where, how often, and how much to win business.

change in scope

An amendment to approved program requirements or specifications after negotiation of a basic contract. It may result in an increase or decrease.

change order/purchase order amendment

A written order directing the seller to make changes according to the provisions of the contract documents.

claim

A demand by one party to contract for something from another party, usually but not necessarily for more money or more time. Claims are usually based on an argument that the party making the demand is entitled to an adjustment by virtue of the contract terms or some violation of those terms by the other party. The word does not imply any disagreement between the parties, although claims often lead to disagreements. This book uses the term *dispute* to refer to disagreements that have become intractable.

clause

A statement of one of the rights and/or obligations of the parties to a contract. A contract consists of a series of clauses.

collaboration software

Automated tools that allow for the real-time exchange of visual information using personal computers.

collateral benefit

The degree to which pursuit of an opportunity will improve the existing skill level or develop new skills that will positively affect other or future business opportunities.

compensable delay

A delay for which the buyer is contractually responsible that excuses the seller's failure to perform and is compensable.

compensatory damages

Damages that will compensate the injured party for the loss sustained and nothing more. They are awarded by the court as the measure of actual loss, and not as punishment for outrageous conduct or to deter future transgressions. Compensatory damages are often referred to as "actual damages." See also *incidental* and *punitive damages.*

competitive intelligence

Information on competitors or competitive teams that is specific to an opportunity.

competitive negotiation

A method of contracting involving a request for proposals that states the buyer's requirements and criteria for evaluation; submission of timely proposals by a maximum number of offerors; discussions with those offerors found to be within the competitive range; and award of a contract to the one offeror whose offer, price, and other consideration factors are most advantageous to the buyer.

condition precedent

A condition that activates a term in a contract.

condition subsequent

A condition that suspends a term in a contract.

conflict of interest

Term used in connection with public officials and fiduciaries and their relationships to matters of private interest or gain to them. Ethical problems connected therewith are covered by statutes in most jurisdictions and by federal statutes on the federal level. A conflict of interest arises when an employee's personal or financial interest conflicts or appears to conflict with his or her official responsibility.

consideration

(1) The thing of value (amount of money or acts to be done or not done) that must change hands between the parties to a contract. (2) The inducement to a contract—the cause, motive, price, or impelling influence that induces a contracting party to enter into a contract.

constructive change

An oral or written act or omission by an authorized or unauthorized agent that is of such a nature that it is construed to have the same effect as a written change order.

contingency

The quality of being contingent or casual; an event that may but does not have to occur; a possibility.

contingent contract

A contract that provides for the possibility of its termination when a specified occurrence takes place or does not take place.

contra proferentem

A legal phrase used in connection with the construction of written documents to the effect that an ambiguous provision is construed most strongly against the person who selected the language.

contract

(1) A relationship between two parties, such as a buyer and seller, that is defined by an agreement about their respective rights and responsibilities. (2) A document that describes such an agreement.

contract administration

The process of ensuring compliance with contractual terms and conditions during contract performance up to contract closeout or termination.

contract closeout

The process of verifying that all administrative matters are concluded on a contract that is otherwise physically complete—in other words, the seller has delivered the required supplies or performed the required services, and the buyer has inspected and accepted the supplies or services.

contract fulfillment

The joint buyer/seller actions taken to successfully perform and administer a contractual agreement and meet or exceed all contract obligations, including effective changes management and timely contract closeout.

contract interpretation

The entire process of determining what the parties agreed to in their bargain. The basic objective of contract interpretation is to determine the intent of the parties. Rules calling for interpretation of the documents against the drafter, and imposing a duty to seek clarification on the drafter, allocate risks of contractual ambiguities by resolving disputes in favor of the party least responsible for the ambiguity.

contract management

The art and science of managing a contractual agreement(s) throughout the contracting process.

contract negotiation

The process of unifying different positions into a unanimous joint decision regarding the buying and selling of products and/or services.

contract negotiation process

A three-phased approach composed of planning, negotiating, and documenting a contractual agreement between two or more parties to buy or sell products and/or services.

contract type

A specific pricing arrangement used for the performance of work under a contract.

contractor

The seller or provider of goods and/or services.

controversy

A litigated question. A civil action or suit may not be instigated unless it is based on a "justifiable" dispute. This term is important in that judicial power of the courts extends only to cases and "controversies."

copyright
A royalty-free, nonexclusive, and irrevocable license to reproduce, translate, publish, use, and dispose of written or recorded material, and to authorize others to do so.

cost
The amount of money expended in acquiring a product or obtaining a service, or the total of acquisition costs plus all expenses related to operating and maintaining an item once acquired.

cost accounting standards
Federal standards designed to provide consistency and coherency in defense and other government contract accounting.

cost contract
The simplest type of cost-reimbursement contract. Governments commonly use this type when contracting with universities and nonprofit organizations for research projects. The contract provides for reimbursing contractually allowable costs, with no allowance given for profit.

cost of goods sold (COGS)
Direct costs of producing finished goods for sale.

cost-plus-a-percentage-of-cost (CPPC) contract
A type of cost-reimbursement contract that provides for a reimbursement of the allowable cost of services performed plus an agreed-on percentage of the estimated cost as profit.

cost-plus-award-fee (CPAF) contract
A type of cost-reimbursement contract with special incentive fee provisions used to motivate excellent contract performance in such areas as quality, timeliness, ingenuity, and cost-effectiveness.

cost-plus-fixed-fee (CPFF) contract
A type of cost-reimbursement contract that provides for the payment of a fixed fee to the contractor. It does not vary with actual costs, but may be adjusted if there are any changes in the work or services to be performed under the contract.

cost-plus-incentive-fee (CPIF) contract
A type of cost-reimbursement contract with provision for a fee that is adjusted by a formula in accordance with the relationship between total allowable costs and target costs.

cost proposal
The instrument required of an offeror for the submission or identification of cost or pricing data by which an offeror submits to the buyer a summary of estimated (or incurred) costs, suitable for detailed review and analysis.

cost-reimbursement (CR) contract
A type of contract that usually includes an estimate of project cost, a provision for reimbursing the seller's expenses, and a provision for paying a fee as profit. CR contracts are often used when there is high uncertainty about costs. They normally also include a limitation on the buyer's cost liability.

cost-sharing contract
A cost-reimbursement contract in which the seller receives no fee and is reimbursed only for an agreed-on portion of its allowable costs.

counteroffer
An offer made in response to an original offer that changes the terms of the original.

customer revenue growth
The increased revenues achieved by keeping a customer for an extended period of time.

customer support costs
Costs expended by a company to provide information and advice concerning purchases.

default termination
The termination of a contract, under the standard default clause, because of a buyer's or seller's failure to perform any of the terms of the contract.

defect

The absence of something necessary for completeness or perfection. A deficiency in something essential to the proper use of a thing. Some structural weakness in a part or component that is responsible for damage.

defect, latent

A defect that existed at the time of acceptance but would not have been discovered by a reasonable inspection.

defect, patent

A defect that can be discovered without undue effort. If the defect was actually known to the buyer at the time of acceptance, it is patent, even though it otherwise might not have been discoverable by a reasonable inspection.

definite-quantity contract

A contractual instrument that provides for a definite quantity of supplies or services to be delivered at some later, unspecified date.

delay, excusable

A contractual provision designed to protect the seller from sanctions for late performance. To the extent that it has been excusably delayed, the seller is protected from default termination or liquidated damages. Examples of excusable delay are acts of God, acts of the government, fire, flood, quarantines, strikes, epidemics, unusually severe weather, and embargoes. See also *forbearance* and *force majeure clause*.

depreciation

Amount of expense charged against earnings by a company to write off the cost of a plant or machine over its useful life, giving consideration to wear and tear, obsolescence, and salvage value.

design specification

(1) A document (including drawings) setting forth the required characteristics of a particular component, part, subsystem, system, or construction item. (2) A purchase description that establishes precise measurements, tolerances, materials, in-process and finished product tests, quality control, inspection requirements, and other specific details of the deliverable.

direct cost
 The costs specifically identifiable with a contract requirement, including but not restricted to costs of material and/or labor directly incorporated into an end item.

direct labor
 All work that is obviously related and specifically and conveniently traceable to specific products.

direct material
 Items, including raw material, purchased parts, and subcontracted items, directly incorporated into an end item, which are identifiable to a contract requirement.

discount rate
 Interest rate used in calculating present value.

discounted cash flow (DCF)
 Combined present value of cash flow and tangible assets minus present value of liabilities.

discounts, allowances and returns
 Price discounts, returned merchandise.

dispute
 A disagreement not settled by mutual consent that could be decided by litigation or arbitration. Also see *claim.*

e-business
 Technology-enabled business that focuses on seamless integration between each business, the company, and its supply partners.

EBITDA
 Earnings before interest, taxes, depreciation and amortization, but after all product/service, sales and overhead (SG&A) costs are accounted for. Sometimes referred to as *operating profit.*

EBITDARM
 Acronym for earnings before interest, taxes, depreciation, amortization., rent and management fees.

e-commerce
A subset of e-business, Internet-based electronic transactions.

electronic data interchange (EDI)
Private networks used for simple data transactions, which are typically batch- processed.

elements of a contract
The items that must be present in a contract if the contract is to be binding, including an offer, acceptance (agreement), consideration, execution by competent parties, and legality of purpose.

enterprise resource planning (ERP)
An electronic framework for integrating all organizational functions, evolved from manufacturing resource planning (MRP).

entire contract
A contract that is considered entire on both sides and cannot be made severable.

e-procurement
Technology-enabled buying and selling of goods and services.

estimate at completion (EAC)
The actual direct costs, plus indirect costs allocable to the contract, plus the estimate of costs (direct or indirect) for authorized work remaining.

estoppel
A rule of law that bars, prevents, and precludes a party from alleging or denying certain facts because of a previous allegation or denial or because of its previous conduct or admission.

ethics
Of or relating to moral action, conduct, motive, or character (such as ethical emotion). Also, treating of moral feelings, duties, or conduct; containing precepts of morality; moral. Professionally right or befitting; conforming to professional standards of conduct.

e-tool

An electronic device, program, system, or software application used to facilitate business.

exculpatory clause

The contract language designed to shift responsibility to the other party. A "no damages for delay" clause would be an example of one used by buyers.

excusable delay

See *delay, excusable.*

executed contract

A contract that is formed and performed at the same time. If performed in part, it is partially executed and partially executory.

executed contract (document)

A written document, signed by both parties and mailed or otherwise furnished to each party, that expresses the requirements, terms, and conditions to be met by both parties in the performance of the contract.

executory contract

A contract that has not yet been fully performed.

express

Something put in writing, for example, "express authority."

express warranty

See *warranty, express.*

fair and reasonable

A subjective evaluation of what each party deems as equitable consideration in areas such as terms and conditions, cost or price, assured quality, timeliness of contract performance, and/or any other areas subject to negotiation.

Federal Acquisition Regulation (FAR)

The government-wide procurement regulation mandated by Congress and issued by the Department of Defense, the General Services Administration, and the National Aeronautics and Space Administration. Effective April 1, 1984, the FAR supersedes both the Defense Acquisition Regulation (DAR) and the Federal Procurement Regulation (FPR). All federal agencies are authorized to issue regulations implementing the FAR.

fee

An agreed-to amount of reimbursement beyond the initial estimate of costs. The term "fee" is used when discussing cost-reimbursement contracts, whereas the term "profit" is used in relation to fixed-price contracts.

firm-fixed-price (FFP) contract

The simplest and most common business pricing arrangement. The seller agrees to supply a quantity of goods or to provide a service for a specified price.

fixed cost

Operating expenses that are incurred to provide facilities and organization that are kept in readiness to do business without regard to actual volumes of production and sales. Examples of fixed costs consist of rent, property tax, and interest expense.

fixed price

A form of pricing that includes a ceiling beyond which the buyer bears no responsibility for payment.

fixed-price incentive (FPI) contract

A type of contract that provides for adjusting profit and establishing the final contract price using a formula based on the relationship of total final negotiated cost to total target cost. The final price is subject to a price ceiling, negotiated at the outset.

fixed-price redeterminable (FPR) contract

A type of fixed-price contract that contains provisions for subsequently negotiated adjustment, in whole or in part, of the initially negotiated base price.

fixed-price with economic price adjustment
A fixed-price contract that permits an element of cost to fluctuate to reflect current market prices.

forbearance
An intentional failure of a party to enforce a contract requirement, usually done for an act of immediate or future consideration from the other party. Sometimes forbearance is referred to as a nonwaiver or as a onetime waiver, but not as a relinquishment of rights.

force majeure clause
Major or irresistible force. Such a contract clause protects the parties in the event that a part of the contract cannot be performed due to causes outside the control of the parties and could not be avoided by exercise of due care. Excusable conditions for nonperformance, such as strikes and acts of God (e.g., typhoons) are contained in this clause.

fraud
An intentional perversion of truth to induce another in reliance upon it to part with something of value belonging to him or her or to surrender a legal right. A false representation of a matter of fact, whether by words or conduct, by false or misleading allegations, or by concealment of that which should have been disclosed, that deceives and is intended to deceive another so that he or she shall act upon it to his or her legal injury. Anything calculated to deceive.

free on board (FOB)
A term used in conjunction with a physical point to determine (a) the responsibility and basis for payment of freight charges and (b) unless otherwise agreed, the point at which title for goods passes to the buyer or consignee. *FOB origin*–The seller places the goods on the conveyance by which they are to be transported. Cost of shipping and risk of loss are borne by the buyer. *FOB destination*–The seller delivers the goods on the seller's conveyance at destination. Cost of shipping and risk of loss are borne by the seller.

functional specification

A purchase description that describes the deliverable in terms of performance characteristics and intended use, including those characteristics that at minimum are necessary to satisfy the intended use.

general and administrative (G&A)

(1) The indirect expenses related to the overall business. Expenses for a company's general and executive offices, executive compensation, staff services, and other miscellaneous support purposes. (2) Any indirect management, financial, or other expense that (a) is not assignable to a program's direct overhead charges for engineering, manufacturing, material, and so on, but (b) is routinely incurred by or allotted to a business unit, and (c) is for the general management and administration of the business as a whole.

General Agreement on Tariffs and Trade (GATT)

A multinational trade agreement signed in 1947 by 23 nations.

generally accepted accounting principles (GAAP)

A term encompassing conventions, rules, and procedures of accounting that are "generally accepted" and have "substantial authoritative support." The GAAP have been developed by agreement on the basis of experience, reason, custom, usage, and to a certain extent, practical necessity, rather than being derived from a formal set of theories.

gross profit margin

Net sales minus cost of goods sold. Also called *gross margin*, *gross profit*, or *gross loss*.

gross profit margin % or ratio

Gross profit margin divided by net sales.

gross sales

Total revenues at invoice value before any discounts or allowances.

horizontal exchange

A marketplace that deals with goods and services that are not specific to one industry.

implied warranty

See *warranty, implied.*

imply

To indirectly convey meaning or intent; to leave the determination of meaning up to the receiver of the communication based on circumstances, general language used, or conduct of those involved.

incidental damages

Any commercially reasonable charges, expenses, or commissions incurred in stopping delivery; in the transportation, care and custody of goods after the buyer's breach; or in connection with the return or resale of the goods or otherwise resulting from the breach.

indefinite-delivery/indefinite-quantity (IDIQ) contract

A type of contract in which the exact date of delivery or the exact quantity, or a combination of both, is not specified at the time the contract is executed; provisions are placed in the contract to later stipulate these elements of the contract.

indemnification clause

A contract clause by which one party engages to secure another against an anticipated loss resulting from an act or forbearance on the part of one of the parties or of some third person.

indemnify

To make good; to compensate; to reimburse a person in case of an anticipated loss.

indirect cost

Any cost not directly identifiable with a specific cost objective but subject to two or more cost objectives.

indirect labor

All work that is not specifically associated with or cannot be practically traced to specific units of output.

intellectual property

The kind of property that results from the fruits of mental labor.

interactive chat

A feature provided by automated tools that allow for users to establish a voice connection between one or more parties and exchange text or graphics via a virtual bulletin board.

Internet

The World Wide Web.

intranet

An organization-specific, internal, secure network.

joint contract

A contract in which the parties bind themselves both individually and as a unit.

liquidated damages

A contract provision providing for the assessment of damages on the seller for its failure to comply with certain performance or delivery requirements of the contract; used when the time of delivery or performance is of such importance that the buyer may reasonably expect to suffer damages if the delivery or performance is delinquent.

mailbox rule

The idea that the acceptance of an offer is effective when deposited in the mail if the envelope is properly addressed.

marketing

Activities that direct the flow of goods and services from the producer to the consumers.

market intelligence

Information on your competitors or competitive teams operating in the marketplace or industry.

market research

The process used to collect and analyze information about an entire market to help determine the most suitable approach to acquiring, distributing, and supporting supplies and services.

memorandum of agreement (MOA)/ memorandum of understanding (MOU)

The documentation of a mutually agreed-to statement of facts, intentions, procedures, and parameters for future actions and matters of coordination. A "memorandum of understanding" may express mutual understanding of an issue without implying commitments by parties to the understanding.

method of procurement

The process used for soliciting offers, evaluating offers, and awarding a contract.

modifications

Any written alterations in the specification, delivery point, rate of delivery, contract period, price, quantity, or other provision of an existing contract, accomplished in accordance with a contract clause; may be unilateral or bilateral.

monopoly

A market structure in which the entire market for a good or service is supplied by a single seller or firm.

monopsony

A market structure in which a single buyer purchases a good or service.

NCMA CMBOK

Definitive descriptions of the elements making up the body of professional knowledge that applies to contract management.

negotiation

A process between buyers and sellers seeking to reach mutual agreement on a matter of common concern through fact-finding, bargaining, and persuasion.

net marketplace

Two-sided exchange where buyers and sellers negotiate prices, usually with a bid-and-ask system, and where prices move both up and down.

net present value (NPV)
The lifetime customer revenue stream discounted by the investment costs and operations costs.

net sales
Gross sales minus discounts, allowances, and returns.

North American Free Trade Agreement (NAFTA)
A trilateral trade and investment agreement between Canada, Mexico, and the United States ratified on January 1, 1994.

novation agreement
A legal instrument executed by (a) the contractor (transferor), (b) the successor in interest (transferee), and (c) the buyer by which, among other things, the transferor guarantees performance of the contract, the transferee assumes all obligations under the contract, and the buyer recognizes the transfer of the contract and related assets.

offer
(1) The manifestation of willingness to enter into a bargain, so made as to justify another person in understanding that his or her assent to that bargain is invited and will conclude it. (2) An unequivocal and intentionally communicated statement of proposed terms made to another party. An offer is presumed revocable unless it specifically states that it is irrevocable. An offer once made will be open for a reasonable period of time and is binding on the offeror unless revoked by the offeror before the other party's acceptance.

oligopoly
A market dominated by a few sellers.

operating expenses
Selling, general, and administrative (SG&A) expenses plus depreciation and amortization.

opportunity
A potential or actual favorable event.

opportunity engagement
> The degree to which your company or your competitors are involved in establishing the customer's requirements.

opportunity profile
> A stage of the capture management life cycle, during which a seller evaluates and describes the opportunity in terms of what it means to your customer, what it means to your company, and what will be required to succeed.

option
> A unilateral right in a contract by which, for a specified time, the buyer may elect to purchase additional quantities of the supplies or services called for in the contract or may elect to extend the period of performance of the contract.

order of precedence
> A solicitation provision that establishes priorities so that contradictions within the solicitation can be resolved.

organizational breakdown structure (OBS)
> An organized structure that represents how individual team members are grouped to complete assigned work tasks.

outsourcing
> A contractual process of obtaining another party to provide goods and/or services that were previously done within an organization.

overhead
> An accounting cost category that typically includes general indirect expenses that are necessary to operate a business but are not directly assignable to a specific good or service produced. Examples include building rent, utilities, salaries of corporate officers, janitorial services, office supplies, and furniture.

overtime
> The time worked by a seller's employee in excess of the employee's normal workweek.

parol evidence

Oral or verbal evidence; in contract law, the evidence drawn from sources exterior to the written instrument.

parol evidence rule

A rule that seeks to preserve the integrity of written agreements by refusing to permit contracting parties to attempt to alter a written contract with evidence of any contradictory prior or contemporaneous oral agreement *(parol* to the contract).

payment

The amount payable under the contract supporting data required to be submitted with invoices, and other payment terms such as time for payment and retention.

payment bond

A bond that secures the appropriate payment of subcontracts for their completed and acceptable goods and/or services.

performance-based contract (PBC)

A documented business arrangement in which the buyer and seller agree to use a performance work statement, performance-based metrics, and a quality assurance plan to ensure that contract requirements are met or exceeded.

performance bond

A bond that secures the performance and fulfillment of all the undertakings, covenants, terms, conditions, and agreements contained in the contract.

performance specification

A purchase description that describes the deliverable in terms of desired operational characteristics. Performance specifications tend to be more restrictive than functional specifications, in that they limit alternatives that the buyer will consider and define separate performance standards for each such alternative.

performance work statement (PWS)

A statement of work expressed in terms of desired performance results, often including specific measurable objectives.

post-bid phase
The period of time after a seller submits a bid/proposal to a buyer through source selection, negotiations, contract formation, contract fulfillment, contract closeout, and follow-on opportunity management.

pre-bid phase
The period of time a seller of goods and/or services uses to identify business opportunities prior to the release of a customer solicitation.

pricing arrangement
An agreed-to basis between contractual parties for the payment of amounts for specified performance; usually expressed in terms of a specific cost-reimbursement or fixed-price arrangement.

prime/prime contractor
The principal seller performing under the contract.

private exchange
A marketplace hosted by a single company inside a company's firewall and used for procurement from among a group of preauthorized sellers.

privity of contract
The legal relationship that exists between the parties to a contract that allows either party to (a) enforce contractual rights against the other party and (b) seek remedy directly from the other party.

procurement
The complete action or process of acquiring or obtaining goods or services using any of several authorized means.

procurement planning
The process of identifying which business needs can be best met by procuring products or services outside the organization.

profit
The net proceeds from selling a product or service when costs are subtracted from revenues. May be positive (profit) or negative (loss).

program management

Planning and execution of multiple projects that are related to one another.

progress payments

An interim payment for delivered work in accordance with contract terms; generally tied to meeting specified performance milestones.

project management

Planning and ensuring the quality, on-time delivery, and cost of a specific set of related activities with a definite beginning and end.

promotion

Publicizing the attributes of the product/service through media and personal contacts and presentations (e.g., technical articles/ presentations, new releases, advertising, and sales calls).

proposal

Normally, a written offer by a seller describing its offering terms. Proposals may be issued in response to a specific request or may be made unilaterally when a seller feels there may be an interest in its offer (which is also known as an *unsolicited proposal*).

proposal evaluation

An assessment of both the proposal and the offeror's ability (as conveyed by the proposal) to successfully accomplish the prospective contract. An agency shall evaluate competitive proposals solely on the factors specified in the solicitation.

protest

A written objection by an interested party to (a) a solicitation or other request by an agency for offers for a contract for the procurement of property or services, (b) the cancellation of the solicitation or other request, (c) an award or proposed award of the contract, or (d) a termination or cancellation of an award of the contract, if the written objection contains an allegation that the termination or cancellation is based in whole or in part on improprieties concerning the award of the contract.

punitive damages
> Those damages awarded to the plaintiff over and above what will barely compensate for his or her loss. Unlike compensatory damages, punitive damages are based on actively different public policy consideration, that of punishing the defendant or of setting an example for similar wrongdoers.

purchasing
> The outright acquisition of items, mostly off-the-shelf or catalog, manufactured outside the buyer's premises.

quality assurance
> The planned and systematic actions necessary to provide adequate confidence that the performed service or supplied goods will serve satisfactorily for the intended and specified purpose.

quotation
> A statement of price, either written or oral, which may include, among other things, a description of the product or service; the terms of sale, delivery, or period of performance; and payment. Such statements are usually issued by sellers at the request of potential buyers.

reasonable cost
> A cost is reasonable if, in its nature and amount, it does not exceed that which would be incurred by a prudent person in the conduct of competitive business.

request for information (RFI)
> A formal invitation to submit general and/or specific information concerning the potential future purchase of goods and/or services.

request for proposals (RFP)
> A formal invitation that contains a scope of work and seeks a formal response (proposal), describing both methodology and compensation, to form the basis of a contract.

request for quotations (RFQ)
> A formal invitation to submit a price for goods and/or services as specified.

request for technical proposals (RFTP)
Solicitation document used in two-step sealed bidding. Normally in letter form, it asks only for technical information; price and cost breakdowns are forbidden.

revenue value
The monetary value of an opportunity.

risk
Exposure or potential of an injury or loss.

sealed-bid procedure
A method of procurement involving the unrestricted solicitation of bids, an opening, and award of a contract to the lowest responsible bidder.

selling, general, and administrative (SG&A) expenses
Administrative costs of running a business.

severable contract
A contract divisible into separate parts. A default of one section does not invalidate the whole contract.

several
A circumstance when more than two parties are involved with the contract.

single source
One source among others in a competitive marketplace that, for justifiable reason, is found to be most worthy to receive a contract award.

small business concerns
A small business is one that is independently owned and operated and is not dominant in its field; a business concern that meets government size standards for its particular industry type.

socioeconomic programs
Programs designed to benefit particular groups. They represent a multitude of program interests and objectives unrelated to procurement objectives. Some examples of these are preferences for small business and for American products,

required sources for specific items, and minimum labor pay levels mandated for contractors.

solicitation

A process through which a buyer requests, bids, quotes, tenders, or proposes orally, in writing, or electronically. Solicitations can take the following forms: request for proposals (RFP), request for quotations (RFQ), request for tenders, invitation to bid (ITB), invitation for bids, and invitation for negotiation.

solicitation planning

The preparation of the documents needed to support a solicitation.

source selection

The process by which the buyer evaluates offers, selects a seller, negotiates terms and conditions, and awards the contract.

Source Selection Advisory Council

A group of people who are appointed by the Source Selection Authority (SSA). The council is responsible for reviewing and approving the source selection plan (SSP) and the solicitation of competitive awards for major and certain less-than-major procurements. The council also determines what proposals are in the competitive range and provides recommendations to the SSA for final selection.

source selection plan (SSP)

The document that describes the selection criteria, the process, and the organization to be used in evaluating proposals for competitively awarded contracts.

specification

A description of the technical requirements for a material, product, or service that includes the criteria for determining that the requirements have been met. There are generally three types of specifications used in contracting: performance, functional, and design.

stakeholders

Individuals who control the resources in a company needed to pursue opportunities or deliver solutions to customers.

standard

A document that establishes engineering and technical limitations and applications of items, materials, processes, methods, designs, and engineering practices. It includes any related criteria deemed essential to achieve the highest practical degree of uniformity in materials or products, or interchangeability of parts used in those products.

standards of conduct

The ethical conduct of personnel involved in the acquisition of goods and services. Within the federal government, business shall be conducted in a manner above reproach and, except as authorized by law or regulation, with complete impartiality and without preferential treatment.

statement of work (SOW)

That portion of a contract describing the actual work to be done by means of specifications or other minimum requirements, quantities, performance date, and a statement of the requisite quality.

statute of limitations

The legislative enactment prescribing the periods within which legal actions may be brought upon certain claims or within which certain rights may be enforced.

stop work order

A request for interim stoppage of work due to nonconformance, funding, or technical considerations.

subcontract

A contract between a buyer and a seller in which a significant part of the supplies or services being obtained is for eventual use in a prime contract.

subcontractor

A seller who enters into a contract with a prime contractor or a subcontractor of the prime contractor.

supplementary agreement

A contract modification that is accomplished by the mutual action of parties.

technical factor

A factor other than price used in evaluating offers for award. Examples include technical excellence, management capability, personnel qualifications, prior experience, past performance, and schedule compliance.

technical leveling

The process of helping a seller bring its proposal up to the level of other proposals through successive rounds of discussion, such as by pointing out weaknesses resulting from the seller's lack of diligence, competence, or inventiveness in preparing the proposal.

technical/management proposal

That part of the offer that describes the seller's approach to meeting the buyer's requirement.

technical transfusion

The disclosure of technical information pertaining to a proposal that results in improvement of a competing proposal. This practice is not allowed in federal government contracting.

term

A part of a contract that addresses a specific subject.

termination

An action taken pursuant to a contract clause in which the buyer unilaterally ends all or part of the work.

terms and conditions (Ts and Cs)

All clauses in a contract, including time of delivery, packing and shipping, applicable standard clauses, and special provisions.

unallowable cost

Any cost that, under the provisions of any pertinent law, regulation, or contract, cannot be included in prices, cost reimbursements, or settlements under a government contract to which it is allocable.

uncompensated overtime

The work that exempt employees perform above and beyond 40 hours per week. Also known as *competitive time, deflated hourly rates, direct allocation of salary costs, discounted hourly rates, extended workweek, full-time accounting,* and *green time.*

Uniform Commercial Code (UCC)
A U.S. model law developed to standardize commercial contracting law among the states. It has been adopted by 49 states (and in significant portions by Louisiana). The UCC comprises articles that deal with specific commercial subject matters, including sales and letters of credit.

unilateral
See *bilateral contract.*

unsolicited proposal
A research or development proposal that is made by a prospective contractor without prior formal or informal solicitation from a purchasing activity.

variable costs
Costs associated with production that change directly with the amount of production (e.g., the direct material or labor required to complete the building or manufacturing of a product).

variance
The difference between projected and actual performance, especially relating to costs.

vertical exchange
A marketplace that is specific to a single industry.

waiver
The voluntary and unilateral relinquishment by a person of a right that he or she has. See also *forbearance.*

warranty
A promise or affirmation given by a seller to a buyer regarding the nature, usefulness, or condition of the goods or services furnished under a contract. Generally, a warranty's purpose is to delineate the rights and obligations for defective goods and services and to foster quality performance.

warranty, express

A written statement arising out of a sale to the consumer of a consumer good, pursuant to which the manufacturer, distributor, or retailer undertakes to preserve or maintain the utility or performance of the consumer good or provide compensation if there is a failure in utility or performance. It is not necessary to the creation of an express warranty that formal words such as "warrant" or "guarantee" be used, or that a specific intention to make a warranty be present.

warranty, implied

A promise arising by operation of law that something that is sold shall be fit for the purpose for which the seller has reason to know that it is required. Types of implied warranties include implied warranty of merchantability, of title, and of wholesomeness.

warranty of fitness

A warranty by the seller that goods sold are suitable for the special purpose of the buyer.

warranty of merchantability

A warranty that goods are fit for the ordinary purposes for which such goods are used and conform to the promises or affirmations of fact made on the container or label.

warranty of title

An express or implied (arising by operation of law) promise that the seller owns the item offered for sale and, therefore, is able to transfer a good title and that the goods, as delivered, are free from any security interest of which the buyer at the time of contracting has no knowledge.

Web portals

A public exchange in which a company or group of companies list products or services for sale or provide other transmission of business information.

win strategy

A collection of messages or points designed to guide the customer's perception of you, your solution, and your competitors.

work breakdown structure (WBS)

A logical, organized decomposition of the work tasks within a given project, typically using a hierarchical numeric coding scheme.

World Trade Organization (WTO)

A multinational legal entity that serves as the champion of fair trade globally, established April 15, 1995.

INDEX